THE SHOCKING
TRUTH ABOUT

by
PAUL C. BRAGG, N.D., Ph.D
LIFE EXTENSION SPECIALIST
and
PATRICIA BRAGG, N
LIFE EXTENSION NU

Health

Happiness   *Youthfulness*

Love   *Joy*

Praise   *Patience*

Vitality   *Fortitude*

Strength   *Charity*

Faith

Eternity   *Salvation*

JOIN
The Bragg Crusades for a 100% Healthy, Better World for All!

HEALTH SCIENCE
Box 7, Santa Barbara, California 93102 U.S.A.

THE SHOCKING TRUTH ABOUT

WATER

By Paul C. Bragg, N.D., Ph.D.
Life Extension Specialist
and
Patricia Bragg, N.D., Ph.D.
Life Extension Nutritionist

— REVISED —

Copyright © Health Science

Twenty-sixth printing MCMLXXXX

ISBN: 0-87790-084-1

Published in the United States by

HEALTH SCIENCE - Box 7, Santa Barbara, California 93102, USA

Fax: (805) 968-1001

WHY THIS BOOK WAS WRITTEN

Paul C. Bragg and daughter, Patricia, bathing in the Miracle Waters of Desert Hot Springs, California. They believe in mineral spas for bathing and swimming therapy but not for drinking purposes.

NEXT TO OXYGEN, WATER IS THE MOST VITAL FACTOR IN THE SURVIVAL OF MAN AND ANIMAL.

Man has survived for as many as ninety days without food, but man can live very few minutes without oxygen and only a few days without water. This is the irony of nature: Man can scarcely exist for 72 hours before going into a semi-comatose state, but it is water, itself that, in most forms, is the very substance which brings about the ultimate ageing of man as well as lesser animals. It not only causes premature ageing, but while living, man can suffer from hundreds of conditions brought on by drinking water saturated with magnesium carbonate, calcium carbonate and other inorganic minerals. This book unlocks the mysteries of why man and animals die long before their time.

Over 50 years in intense research has gone into this book.

YOURS FOR HEALTH AND LONG LIFE,

Patricia Bragg   *Paul C. Bragg*

"The Noblest Of The Elements Is Water." — Pindar

Paul C. Bragg, N.D., Ph.D.
"World's leading authority on pure water!"

PAUL C. BRAGG SAYS, "YOU SHOULD KNOW THE SHOCKING FACTS ABOUT WATER BEFORE YOU DRINK IT." IT'S SERIOUS BUSINESS.

CONTENTS

VI

INTRODUCTION

PURE WATER (H²O) A PRIME REQUISITE OF HEALTH

The most important factors that make for a long healthy life are:

- Pure Air.
- Absolutely Pure Water free from all chemicals and inorganic minerals.
- Pure Food as natural as you can get it today in this poisoned and polluted world.

The most important substance on earth is — WATER! Water is one of the most characteristic substances of our planet. It may simultaneously appear in solid, liquid and gaseous forms. It has been adapted as a unit of measure for the specific gravity of all other substances. It plays an important role in the circulation of the elements on the earth's surface.

Man must have water, or he soon dies! Take the ship-wrecked sailor on a great ocean of salt water: If this man does not get fresh unsalted water, he dies. The man who gets lost in the torrid, hot desert soon dies of dehydration if he does not get water. Thirst can drive him insane before he dies an agonizing death.

Certain animals, such as hares and rabbits, which feed on grasses and herbs containing about 85% water, never drink so long as they can find their natural food. Mother's milk contains about 87% water; juicy fruits and succulent vegetables also possess almost the same fluidic content. The person who consumes about four pounds of fresh fruit daily absorbs, in addition to about eight ounces of solid food, at least three pints of live, living naturally purified-by-nature water.

In my opinion, water is the most important substance on the face of the earth. Without it, life — from plant to human — would cease to exist.

VII

WATER GOES ON FOREVER

Water is absolutely indestructible. Scientists believe that there is not a drop more — nor a drop less — than when shallow water first formed the roundness of the earth with its tidal currents. Volcanic eruptions eventually brought solid rock and earth above the water in the form of mountains. In eons of time, these became continents.

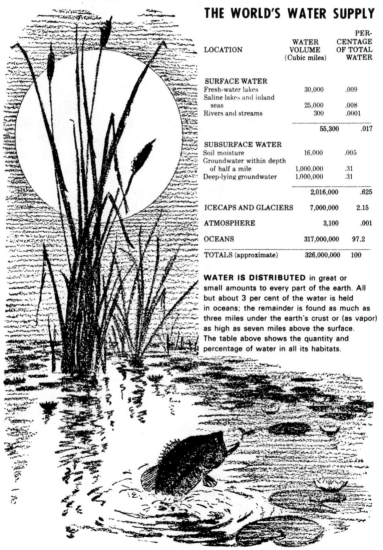

THE WORLD'S WATER SUPPLY

LOCATION	WATER VOLUME (Cubic miles)	PERCENTAGE OF TOTAL WATER
SURFACE WATER		
Fresh-water lakes	30,000	.009
Saline lakes and inland seas	25,000	.008
Rivers and streams	300	.0001
	55,300	.017
SUBSURFACE WATER		
Soil moisture	16,000	.005
Groundwater within depth of half a mile	1,000,000	.31
Deep-lying groundwater	1,000,000	.31
	2,016,000	.625
ICECAPS AND GLACIERS	7,000,000	2.15
ATMOSPHERE	3,100	.001
OCEANS	317,000,000	97.2
TOTALS (approximate)	326,000,000	100

WATER IS DISTRIBUTED in great or small amounts to every part of the earth. All but about 3 per cent of the water is held in oceans; the remainder is found as much as three miles under the earth's crust or (as vapor) as high as seven miles above the surface. The table above shows the quantity and percentage of water in all its habitats.

VIII

These tides, conversely, slow the rotation of the earth by a fraction of a second every thousand years. The 24-hour day was possibly a 4-hour day a billion years ago. Originally, the earth probably consisted of hot gases. As it cooled, hydrogen and oxygen atoms fused and formed a steamy mist. Much later, the steamy mist fell in torrential endless rains, the coolness of which formed a "solid" floor.

Water shapes the earth, controls the climates, provides man with food and a prodigious amount of energy. Water constitutes four-fifths of the body weight and performs and supports the internal functions of animals and plants.

THE WONDERS OF GOD-GIVEN WATER ARE ENDLESS

It is possible that a tear, which fell from the eye of Jesus when he learned his friend Lazarus had died, has been recycled by the warmth of the sun millions of times and may repose in a holy water fountain in some obscure church.

WATER PENETRATES EVERYWHERE

The molecular strength of a drop of water is almost beyond

The Hydrologic Cycle

understanding. As it penetrates the lacy roots of a big tree, it starts to climb upward, pulling after it a chain of water drops.

The wind will vaporize the water in the topmost leaf of the tree, carrying it back to the sky to help form a rain-bearing cloud. The same drop may be carried as much as seven miles above the earth, to remain wind-borne and purified before dropping with billions of others, perhaps on an orchard of apples. Or the raindrop may be caught as rain by a group of shipwrecked sailors on a waterless island. It may fall on the parched ground of the Indians of Arizona and bring back life to a seed that needs only water to grow. An inch of rain on a square mile of topsoil comes to over 17 million gallons of water.

One such raindrop, if it lingers on the surface of the earth, may be revaporized and head for the sky in less than a minute. If it penetrates deeply into the water table far below the Sahara desert, where 150,000 cubic miles of water stand waiting, it may require a century to resurface and become air-borne.

HOT MINERAL WATER FLOWS UNDER CALIFORNIA DESERT

Just a few hundred feet below the desert in California where I have a home, there is a raging river of hot mineral water. Wells are sunk down to reach this water, which comes out at a temperature as high as 180 degrees. This water has been underground for centuries. The water is cooled to temperatures that human beings can tolerate and it gives blessed relief to thousands who have aches and pains all over their bodies. This warm mineral water is so very relaxing and therapeutic, people come from all over the world to bathe in its healing warmth.

I have a painless body, yet I still take the time (preventive medicine) to enjoy these hot mineral baths for they are soothing and relaxing to my body. This is the reason I built a home in this Spa town; so when I want to relax and be quiet away from big cities, this is where I go — to my desert retreat home.

We have Angel View Crippled Children's Hospital in our town, and it is world-famous for the miracles it is doing with crippled children. It does my heart good everytime I go up to visit the children, to see them swimming in their hot mineral therapeutic swimming pool! Even though many of the children cannot walk when they arrive, they soon learn how to swim in the pool and this starts to build confidence into their little bodies! This water therapy, plus physical therapy, has worked miracles under the guidance of my good friend, Dr. Frank Edmundson, M.D. Our late President, Dwight D. Eisenhower, was on the Board of Directors of this hospital for many years.

The oceans hold 97.2 per cent of all water, which is used in enormous tides, waves and winds, to crash and slam the rocky beaches and reduce them to sand. Where volcanic eruptions have flowed to the sea, this volcanic material in time is reduced to sand, that is why the Isles of Hawaii and Tahiti have black sandy beaches. The sea will always win the battle with the earth. Geologists tell us that eventually the highest mountain will be levelled under the ocean and the cycle of volcanic eruptions will begin again.

Eons of years ago, glaciers were so numerous that the level of the seas fell 300 feet and land bridges between England and France appeared. This also occurred between Siberia and Alaska. That may account for some of the mystifying migrations of peoples.

NIAGARA FALLS COMMITTING SLOW SUICIDE

A short time ago, engineers diverted the water of Niagara Falls to study a means of stopping the erosion of slag. Man cannot stop it. Niagara drops 3,500,000 gallons of water over the edge of the falls every second. In a little more than 20,000 years, the falls will retreat to Lake Erie and become a level river.

The famed oceanographer, Columbus Iselin, chided science when he wrote: "The sea is producing about as much as the land, yet man is taking only about one per cent of his food from his salt water environment." Man is more interested in the unknown darkness of outer space than studying the richness of the seas!

A solitary drop of water is a strange world indeed.

MAN CANNOT LIVE WITHOUT WATER

Your existence on earth depends on WATER! Please do not take it for granted. This book gives you an education on the values of the perfect water to drink that will help to keep you in good health and also help you prolong a more vital, joyous living on the top of this good earth every day of your life.

THE FIVE BIG HEALTH BUILDERS
• AIR • WATER • SUNSHINE • FOOD • EXERCISE

Next to oxygen, water is the most important substance in the body. The average adult body contains approximately 45 quarts of water and loses about three and a half quarts daily in perspiration, respiration, urine and defecation. The temperature of the body is controlled through water. The average body temperature is 98.6. If it goes over this, we are feverish; and if it goes under this, we are under par physically. Water makes up 92 per cent of the blood of the body and nearly 98 per cent of intestinal, gastric, saliva and pancreatic juices. Many,

many people have that dehydrated look. Their skin looks parched, withered, dry, and old. Look at the dry and withered hands of many people. Look at the wrinkles on the forehead and around the eyes. See how a curtain of dry flesh hangs over the eyes. Such people seem to squint out of little slits to see.

Many people are chronically constipated which is another sign of dehydration. Many people suffer from burning, irritating urine which is another sign of water starvation.

SALT — A HEALTH WRECKER

Then there is the other side of water imbalance due to the excess use of salt and salty foods, the water-logged human. You see young children — seven, eight, nine and ten years and older who are so water-logged they look grotesque. Some of these boys and girls have bloated and middle-aged looking bodies. Look at the adults with bloated moon-faces and puffed-up arms, bellies, legs, ankles and feet.

The amount of water the body needs depends on temperature, climate, one's activities, and general health. When you drink a glass of water, it goes directly to your stomach. Part of the water is absorbed directly into your blood stream through the walls of the stomach. The remainder goes to the intestines to keep the food you eat in a liquid state while being absorbed; this is later absorbed directly into the blood.

The right kind of water is one of your best natural protections against all kinds of virus infections, such as influenza, pneumonia, whooping cough, measles and other infectious diseases.

During the Hong Kong flu, doctors advise bed-rest and plenty of water. When the body tissues and cells are kept supplied with the correct amount of water, they can fight off the attacks of viruses. If the body cells are water-starved, they become parched, dry, and shrivel, making it easy for viruses to attack.

Bear in mind important functions of the right kind of water in your body: Water is a vital factor in all body fluids, tissues, cells, lymph, blood and all glandular secretions; water holds all nutritive factors in solution and acts as a transportation medium to various parts of the body; it holds toxins and body wastes in solution and again serves as a transportation medium of these substances; the mucous membranes need plenty of water to keep them soft and free from friction on their delicate surfaces. Liquid is necessary for the proper digestion of food. The stomach acts as a powerful churn in breaking down food into small particles.

WATER FLUSHES BODY TOXINS OUT

One of the most important functions of water is to flush the toxins and salt from the body. Unfortunately people the world over use large amounts of salt. Centuries ago to the present day, whole countries never know what salt is and still are healthy and happy.

The Japanese are known to be the world's highest consumers of salt (sodium chloride). A Japanese farmer who lives to be sixty years of age eats approximately two ounces of salt every day of his life and filters through his kidneys 2,737.5 pounds or 1.36 tons of salt in his lifetime.

Americans are not far behind the Japanese in the consumption of salt. Not only do the Americans shake plenty of salt on their foods, they also eat large amounts of salty foods, such as ham, bacon, hot dogs, luncheon meats, corned beef, potato chips, salted nuts, and many other foods with high concentrations of salt. No wonder heart disease is the No. 1 killer in America! No wonder people in their thirties suffer from high blood pressure, kidney trouble, arthritis, and the beginning of the greatest of all killers, hardening of the arteries, veins and blood vessels!

The proper amount and right kind of water helps to keep the cholesterol level down. Remember water is a flushing agent. In my opinion, the right kind of water is nature's greatest beauty and health tonic. In my long career as a nutritional and conditioner advisor to the greatest stars in films and TV, I have found that, when I can persuade the stars to use the correct liquids, they retain the ageless face, form and figure very much longer than the person who drinks ordinary water. Correct water keeps the body cells normal and prevents dehydration. The face and neck are more free of ageing lines and wrinkles and the face and body retain the characteristics of youth longer.

WATER IMPORTANT TO SUPERB HEALTH

People who drink the right kind and the right amount of liquids (distilled water, fresh fruits and vegetables and their juices) have better circulation which is most important to super-health and long life.

In my personal opinion, I think the right kind of water helps improve your mind and brain power, and I really think it makes you think more accurately and better. You have 15 billion powerful brain cells which are 70% water.

I further think that the excessively nervous person and/or the mentally upset person, is so obsessed with his own worries and "hang-up" that he just forgets to drink water and liquids of the right kind. Instead he dopes himself on alcohol, tea, coffee, and cola drinks and thus complicates his nervous condition by burning, toxic acid forms in the stomach with no food or water to dilute it. So on top of his nervousness and depression, he suffers from sour acid stomach, heartburn, gas-bloat and other miseries. Instead of drinking sufficient water, these people dope up on aspirin, cigarettes and other stimulants.

Remember that the nerves need the correct amount of water to function properly and smoothly. You can plainly see that it is possible to suffer from water starvation. Here is a way you can help yourself to health — the Natural Liquid Way.

That is the specific reason I wrote this book so you will have the knowledge to select the right kind and the right amount of liquid your body so desperately needs.

Here is your invitation to enjoy the natural gift of Super-Health — Health and Freedom from body miseries through powerful self-knowledge.

THE 65% WATERY HUMAN

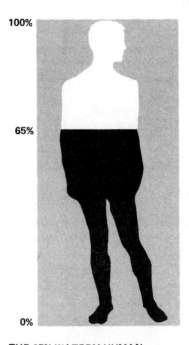

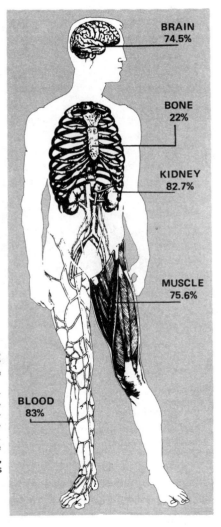

BRAIN
74.5%

BONE
22%

KIDNEY
82.7%

MUSCLE
75.6%

BLOOD
83%

THE 65% WATERY HUMAN

The amount of water in the human body, averaging 65 per cent, varies considerably from person to person and even from one part of the body to another (right). A lean man may have as much as 70 per cent of his weight in the form of body water, while a woman, because of her larger proportion of water-poor fatty tissues, may be only 52 per cent water. The lowering of the water content in the blood is what triggers the hypothalamus, the brain's thirst center to send out its familiar demand for a drink.

"Water Is The Best Drink For A Wise Man" — Henry Thoreau

PATRICIA BRAGG, Life Extension Nutritionist, enjoys her home in Santa Barbara, California, where she has organic gardens . . . including even delicious fruit-bearing banana trees.

Important Message for Readers

WHAT THIS AMAZING BOOK CAN DO FOR YOU

A HEALTH MESSAGE 100 YEARS AHEAD OF OUR TIME.
A miraculous transformation occurs within you when you discover what is the perfect drinking water of Man.

Paul C. Bragg sincerely believes that he's discovered the World's Most Important Health Secret. For more than fifty years he has been researching the subjects of illness and ageing. At last he feels he has found the answer as to why man becomes sick, ages prematurely and dies before his time. Read this remarkable book and learn why this work is of greatest interest for at least half of our population.

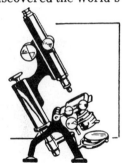

ASK YOURSELF THESE QUESTIONS:

- How can I stop the chemicals and inorganic minerals from turning my brain and body into stone?
- How can I stop my joints from becoming stiff and cemented?
- How can I help stop the formation of gall, kidney and bladder stones?
- How can I keep my arteries, veins and capillaries from this unnatural hardening?
- How can I prolong Youth?
- How can I delay the onset of premature ageing?

To Find The Answers — Read This Book!

DRINKING WATER OFFERS DEADLY CHEMICALS!

With the scrupulous precision of a responsible scientist, Paul C. Bragg describes the serious, devastating danger in drinking ordinary water loaded with deadly commercial chemicals and inorganic minerals, plus the use of common table salt.

Bragg tells why the chlorination of our public water supplies may not be the innocent thing that it appears to be. Just remember that chlorine - nascent and the hypochlorite, chlorine dioxide and other chlorine compounds - are strong oxidizing and bleaching agents. When the chlorination of our drinking water is sufficient to produce an offensive taste and smell, enough chlorine may enter the intestinal tract to destroy helpful bacteria and thereby deprive us of the important vitamins which they make for us.

Bragg tells you about artificially fluoridated drinking water which more than 80 million human guinea pigs are now drinking in an unprecedented experiment which less venturesome nations are watching with alarm and amazement.

Bragg tells how inorganic salt may cause many troubles in the body.

Bragg believes he has found the reason we have more hospitals, nurses, mental institutions, medical doctors and other healers and more medical colleges than ever before in history.

Bragg believes he has found the reason why more people are dying of degenerative diseases, such as heart trouble, arthritis, kidney trouble and hardening of the arteries long before their time. And to top it all, we have a greater number of retarded, deformed, crippled children than ever before in history.

From start to finish, this book is enthralling...a quietly terrifying and immensely serviceable book. It gives, moreover, admirable advice. This book will be a source of positive and practical enlightenment for everyone who is vitally interested in the problems of regaining and maintaining good health.

It shows the way to be healthier, more youthful, happier, and how to help add energetic years to your life.

Read this book and discover some revealing facts about yourself, your health and your chances of enjoying life for many years to come.

THE SHOCKING TRUTH ABOUT WATER

"Water! Water! Everywhere — but not a drop to drink!"

Yes, with all the billions of gallons of fresh, sweet water there is on earth, little of it is fit to drink. Water, a chemical compound having the formula of H_2O, is one of the most abundant and widely distributed substances on the surface of the earth. It occurs in nature in solid, liquid and gaseous states of aggregation known as ice and snow, water, and steam vapor, respectively. Water, composed of hydrogen and oxygen, is contained in varying amounts in all natural foods. It is indispensable as a solvent in all of the physiological functions of the body and in every form of life.

The "Black Death" that spread throughout Europe in the 1300's killed one-third of the entire population. This plague was caused by polluted water. Even today in many parts of Europe, the water is unfit to drink. People use bottled water for drinking purposes. Every experienced world traveler drinks bottled water.

The body requires water — but bear in mind that it must have chemically pure water, water that is 100% pure hydrogen and oxygen. This pure water comes from two sources: first, from fresh organically-grown vegetables and fruits. (Nature naturally purifies the water in healthy organically-grown vegetables and fruits.) Second, from distilled water made by the steam process or by one of the new high-efficiency deionization processes.

Much of our water today is polluted. It is difficult to find sources

1

of water from rivers, lakes, streams, and, now, even wells and springs which are not to some degree polluted. Therefore, a great deal of chlorinization is necessary to make this water fit to drink.

But is it really "fit to drink" even then? Remember, the water-processing plants use inorganic chlorine to fight the bacteria of polluted water. Alum and many other inorganic chemicals are also used to cleanse polluted water of dirt and filth.

On top of these inorganic chemicals, the worst inorganic substance has been added to drinking water—that is, inorganic sodium fluoride. In my opinion, this is the worst thing that has ever happened to drinking water.

INORGANIC VS. ORGANIC MINERALS

Now, let me give you a short lesson in chemistry. There are two kinds of chemicals, inorganic and organic.

The inorganic chemicals like chlorine, alum and sodium flouride cannot be healthfully utilized by the living tissues of the body.

Our body chemistry is composed of 19 organic minerals, which must come from that which is living or has lived. When we eat an apple or any other fruit or vegetable, that substance is living. It has a certain length of life after it has been picked from the vine or tree. The same goes for animal foods, fish, milk, cheese and eggs.

Organic minerals are vital in keeping us alive and healthy. If we

were cast away on an uninhabited island where nothing was growing, we would starve to death. Even though the soil beneath our feet contains 16 inorganic minerals, our bodies cannot absorb them efficiently enough to sustain life. Only the living plant has the power to extract inorganic minerals from the earth.

When I was on an expedition in China many years ago, one part of the country was suffering from drought and famine. With my own eyes, I saw poor, starving people heating earth and eating it for the want of food. They died horrible deaths because they could not get one bit of nourishment from the inorganic minerals of the earth.

For years I have heard people say that certain waters were "rich in all the minerals." What kind of  minerals are they talking about? Inorganic or organic? Humans do not have the same chemistry that a plant does. To repeat, only a living plant can convert an inorganic mineral into an organic mineral. And as you read this book, you will learn what harm inorganic minerals can do to your body and your brain.

Because of a dietary lack, some children and young animals eat dirt. They become deathly sick — not from the germs which were in the dirt, but from the inorganic minerals which can cause illness and even death.

During my boyhood on a Virginia farm, we raised dairy cattle. Sales-  men would come to our farm to sell various kinds of feed for these animals. I well remember when my father purchased cattle food labeled "The Mighty Mineral Cattle Food." It was supposed to have lots of calcium, magnesium and other important minerals to help build strong cows that would produce extra milk; however, all the minerals in this "mighty" cattle food were from inorganic sources: calcium carbonate, magnesium carbonate, etc., mostly from powdered limestone. When this inorganic mineral formula was mixed with organic food, our cattle absolutely refused to eat the stuff. Their innate natural instinct prevented them from eating powdered limestone. Our farm neighbors had the same experience. We learned later that this inorganic mineral cattle food was taken off the market as worthless.

DANGEROUS INORGANIC MINERALS
IN DRINKING WATER

As previously noted, chlorine, alum and other inorganic minerals are put into our drinking water for cleanliness. In addition, other dangerous inorganic minerals are used, such as calcium carbonate and its chemical affinities, magnesium carbonate, potassium carbonate and others.

Keep constantly in mind that **the human body needs only hydrogen and oxygen as a natural solvent in the body chemistry.**

This wild animal of the deep African jungle, if not killed in battle with another animal will be killed by drinking the hard water found in Africa.

Yet the body must have a constant supply of water. Where to get it? Even untreated, so-called "pure" water — from springs, wells, etc. — nearly always contains some inorganic minerals, as will be discussed later.

This is the irony of Nature: that this fluid — without which man can barely exist more than 74 hours without passing into a semi-comatose state — contains in itself, in most forms, the exact inorganic chemicals which bring about the ultimate premature ageing of man as well as lesser animals.

And, as stated earlier, NOW the great aluminum companies want to drug all our water with sodium fluoride, a worthless waste produced by aluminum processing.

FLUORINE IS A DEADLY POISON

Millions upon millions of innocent people have been brainwashed by the aluminum companies to believe that by adding sodium fluoride to our drinking water we will reduce tooth decay in our children. Today, without a thought, nearly 80 million Americans drink a daily dose of sodium fluoride with their drinking water.

Many honest and sincere chemists say fluoridation is the unbelievable blunder.

Fluorine, the gangster of the chemical underworld, made the atomic bomb possible. The only scientific way to free the quantities of fissionable Uranium 235, buried in the inert mass of its parent U-238, was to force uranium hexafluoride gas through many acres of porous barriers, gradually concentrating the precious element. "Hex," they named the vicious stuff — and "Hex" took its deadly toll, rotting out barriers, eating piping and pumps, creating a terrible hazard from radation.

Millions of Americans drink water with sodium fluoride in solution. Connected to sodium, fluorine is not as violent as "Hex"

— but violent enough in high concentrations to be used as a standard roach and rat killer and a number one pesticide.

Yet this terrible fluoride, inserted virtually by government edict into drinking water in the proportion of 1.2 parts per million (PPM), is declared by the U.S. Public Health Service to be "absolutely safe for human consumption". Every qualified chemist knows that such "absolute safety" is illusory.

THE GRIM STORY BEHIND FLUORIDATION

It was in the year 1939 that a famous institute in the eastern part of the United States commissioned the institute's biochemist to find a use for the sodium fluoride wastes produced by aluminum pot lines. Some 45 other industries had fluoride disposal problems, too. Many were tormented by expensive damage suits arising from the noxious effects of the poison on livestock and crops. Brick, steel, fertilizer, oil refineries, metal smelters, tile, ceramics plants, and many installations of the Atomic Energy Commission were involved. The cost of elimination of the chemical was fantastically high. Couldn't this by-product be put to a profitable use instead?

Now, this biochemist was a clever and cunning man, and he came up with a big money-making idea: Why not dissolve the stuff in drinking water? Now, this biochemist had absolutely no medical background and had made not one clinical research on the action of sodium fluoride in the body chemistry. His idea went over big with the companies who were bedeviled with the sodium fluoride wastes.

The next step was not hard. Turn the idea over to the advertising companies and let them brainwash the public mind that the greatest health measure in modern times had been discovered. Give a "sob story" to the American public. The American public is so very gullible. If it is scientific sounding, the public will bite — hook, line and sinker. So they used the tall tale that sodium fluoride in drinking water would prevent tooth decay in children. That line of sales talk hit a responsive chord. At last,

---

Surely if living creatures saw the consequence of all their evil deeds, with hatred would they turn and leave them, fearing the ruin following.

—F'shuing Tsan K'ung

"I have found a perfect health, a new state of existence, a feeling of purity and happiness, something unknown to humans . . . "

—Novelist Upton Sinclair,
who fasted frequently.

6

a way to prevent tooth decay! Every intelligent and thinking person knows that tooth decay comes from poor nutrition, especially the use of refined white sugar.

Poor nutrition, refined sugars and their products, soft drinks, sweets, etc., are causing epidemic tooth decay among Americans.

Big business and big professional organizations have a way of sticking together. And remember, they have most of the news media with them because of the economic control exerted by the publishers' chief source of income, advertising. So the great business and professional organizations — with the aid of newspapers, magazines, radio and television — blasted out the sodium fluoride brainwash to the American people. Any person who questioned the poisoning of drinking water with sodium fluoride was called stupid, dishonest, ignorant, uninformed and backward! Most doctors and dentists surrendered to these powerful forces for fear of being discredited in their communities although you can always find some honest and sincere professional men who fight for truth — no matter how others might ridicule them! Extreme pressures were put upon city and state governments to fluoridate the drinking water. Big business and big professional organizations, which can act like the Mafia, do not take "No" for an answer! They know how to make state and city officials think their way.*

DON'T DRINK WATER CONTAMINATED
WITH SODIUM FLUORIDE

Fluorine is one of the most potent poisons known to man. Selling of this poison is big business, and it swells the bank accounts of the big companies who in turn pay big dividends to their stockholders. And to think that this money

"Money is the root of all evil."

is made by selling a waste by-product! All these companies had to do was to brainwash the public mind to accept their statement that fluorine in drinking water would prevent tooth decay in children, use the national news media for propaganda, and have strong lobbies on all levels of government — national, state and city.

But things are different now. City drinking water is being rapidly fluoridated due to the powerful organizations who are sponsoring this mass poisoning. In my opinion, many early deaths today are due to arteries that have been prematurely aged by sodium fluoride. Not only arteries have been damaged, but also the heart, lungs, liver and other vital organs.

No person suffered from the dangers of fluorine (sodium fluoride) poisoning 100 years ago, nor did any animal. It wasn't in use!

* **Documentary "The Grim Truth about Fluoridation", by Peter Gray**

USA PER CAPITA DAILY HOME CONSUMPTION IS 60 GALLONS PER PERSON OR 14 BILLION GALLONS DAILY

WE LIVE IN A SICK, SICK WORLD

Read this book carefully. More than 70 years of tedious research work has gone into acquiring the information given here.

I am an independent researcher for the truth. I am a loner. No one controls what I say. No organization dictates to me. I have no master to serve; therefore I can give you the plain, honest truth.

As author of this book I'm a man of science — the science of helping the total man to health and a more vibrant youthful life through natural living! I spend a large part of my time in a laboratory studying living matter, and an equal amount of time in the outside world watching and studying human beings, trying to understand why they become sick, prematurely old, senile, and why they die long before their time.

It is my conclusion that man makes himself sick and shortens his life with nature's universal solvent, water — plus tobacco, powerful stimulants such as alcohol, coffee, tea, cola drinks, soft drinks of all kinds; by high concentrations of refined white sugar and its products, refined white flour and its products, white rice, salt and salted foods; by overeating and by eating dead, devitalized, demineralized foods, devitamized foods, too many saturated fatty foods (which fill the arteries and other blood vessels with waxy, gooey cholesterol).

Food can make or break you. You can dig your grave with your knife, fork and spoon.

Whatsoever was the father of a disease; an ill diet was the mother.
— Herbert, 1859

These bad habits — plus lack of exercise, sunlight and fresh air — add up to physical **unfitness.** 97% of the people today are physically unfit. 65% are victims of overweight. Most people eat such an incorrect diet that they suffer from extreme fatigue. They drag themselves around all day and at night are forced to take a sleeping pill to get to sleep. When they awaken, they take a "pep pill" to get them going.

We are a nation of pill takers. Every 24 hours between 30 and 35 tons of aspirin is consumed.

Physical fatigue and weariness make them more inactive than ever. They just do not have the "Go Power" to lead active physical lives. As a consequence, their exterior as well as interior muscles suffer from increasing flabbiness. The greatest "disease" today is the physical deterioration of the human body.

Let's take a good hard look at our young people. Never in the entire history of this country have so many drugs been used by people under thirty. Why do young people need powerful drugs to keep them going? Just take a look at the "junk"

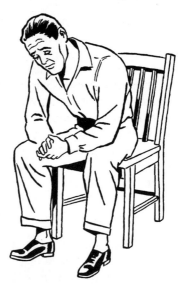

This man is a victim of chronic fatigue. He has no energy, vitality, strength or ambition. He suffers from extreme weariness physically and mentally, a state of physical deterioration.

and "trash" they eat — that tells the entire story! They do not derive the proper nutrition from their daily diet, and in their

10

ignorance, try to supply the necessary energy by harmful stimulants such as tobacco, alcohol and drugs. And these "dope addicts" are to be the parents of future Americans!

I can only repeat, "We live in a sick, sick world, and it is getting sicker every day."

FEW PEOPLE KNOW THE MEANING OF HEALTH

To many people health is something they value only when they have lost it or are in danger of losing it.

"Health" is an old Anglo-Saxon word meaning "Soundness". The concept of "a sound mind in a sound body" (**mens sana in corpore sana**) gives the honest picture of health. We have to put the adjective "good" with this world "health" only when we contrast it with the phrase "ill health" or "lack of soundness".

A healthy, vigorous, energetic body and an alert, keen, healthy mind make it possible for us human beings to carry the frustrations, worries, cares, tensions, stresses and strains — and share the joys — of this world, even as mixed-up, confused and feverish as it is today. Where there is vigorous health, there is not even an awareness of the complex mechanisms and chemistry which go on within us to make this possible.

We take our health for granted, as we do the moon and the sun, usually even neglecting to be thankful for it. We get up in the morning after a sound night's sleep, ready to take on the day's work, and to end the day comfortably tired.

Within these bodies of ours, the most magnificent chemical and mechanical procedures known to man — and some still unknown — are carried on.

THE MARVELOUS MECHANISM OF THE BODY

If we had transparent fronts and could look inside each morning, we would see the lungs taking air into their delicately fashioned units and cells. And if we were smokers, we could see the vicious nicotine and tars turning these beautiful, pink, healthy organs to sticky, dirty black.

We could see the heart receiving blood by intricate channels from the billions of body cells and pumping it out, refreshed and purified by another route to these same cells. We would get an exact picture of our arteries, veins and capillaries. We could see

11

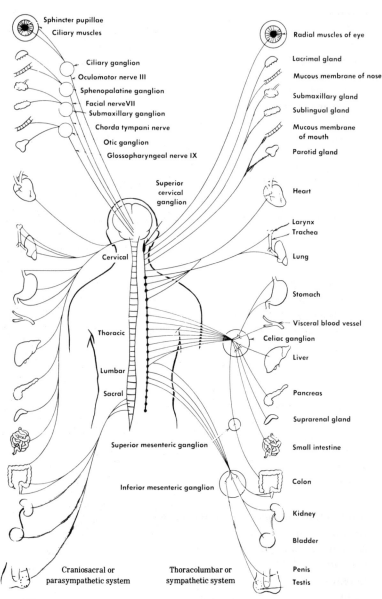

The autonomic nervous system showing its two divisions: the craniosacral or parasympathetic, and the sympathetic.

"The nervous system suffers when we do not take care of our body."

how much corrosion is taking place because of our consumption of the heavy inorganic chemicals that are put into our drinking water. If we examined our arteries closely, we could see that calcium carbonate and its affinities are lining these pipes and making them brittle — beginning to turn our body into stone. Oh, if we could see what inorganic chemicals do to our arteries, we would certainly take the advice given in this book! Remember, we are as young as our arteries.

If we could look inside ourselves, we could also see the digestive tract performing miraculous changes in the food and drink we give it — thus making it possible for our body cells to use salads, nuts, seeds, raw and cooked vegetables, and other proper nourishment which is transformed into the substances which these cells demand and can use. And the person who lives on a dead, de-

Sodium nitrate is one of the food additives that's harmful.

vitalized diet would see how the body chemistry struggles to handle hot dog sandwiches, commercial ice cream, doughnuts, and all other "food trash" which insults their digestive tracts.

If we could get a full view of the largest organ in the body, the liver, we would see how it struggles to handle alcohol, coffee, tea, cola drinks and improper foods. We could see the disastrous effects of the powerful inorganic chemicals that are put into our drinking water by man — and those put there by nature. For nature can sometimes outdo man by contaminating water with inorganic minerals such as calcium carbonate, magnesium carbonate, potassium carbonate, and many others. Looking closely, we could see that the liver is hardening, turning to stone.

Thousands upon thousands of people die from a disease known as cirrhosis of the liver — fibrosis with hardening caused by excessive formation of connective tissue followed by contraction of the liver. Hard water inorganic minerals can harden the liver.

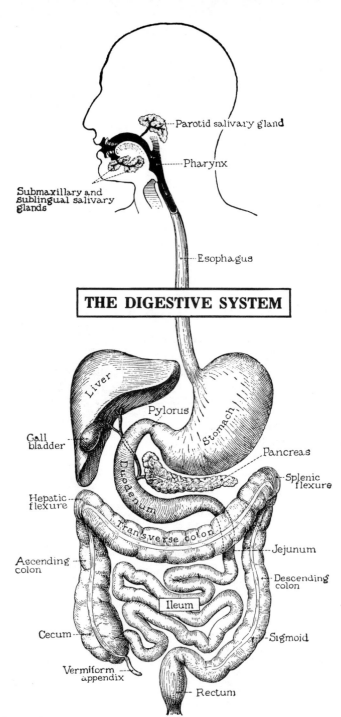

Parotid salivary gland

Pharynx

Submaxillary and
sublingual salivary
glands

Esophagus

THE DIGESTIVE SYSTEM

Liver

Pylorus

Stomach

Gall
bladder

Pancreas

Duodenum

Splenic
flexure

Hepatic
flexure

Transverse colon

Jejunum

Ascending
colon

Descending
colon

Ileum

Cecum

Sigmoid

Vermiform
appendix

Rectum

14

HARDENING OF THE ARTERIES

On several occasions during my boyhood in Virginia, my parents took me to the famous Luray Limestone Caverns. There I saw how, drop by drop, water loaded with limestone formed the stalactites and stalagmites through eons of time. These were huge formations of inorganic minerals that had been built slowly, one drop at a time. This is exactly the same process that manifests itself inside our bodies in the form of calcium carbonate and other inorganic minerals that are ever-present in our drinking water.

Calcium carbonate, or lime, is the important ingredient in making cement or concrete.

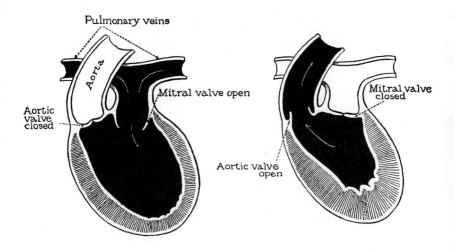

THE ARTERIES AND VALVES OF THE HEART MUST BE FREE FROM DEPOSITS OF INORGANIC MINERALS AND TOXIC CRYSTALS.

Diastole and Systole in the left heart. Note the positions of the two valves in both cases. Diastole to the left, Systole to the right. You are as young as your arteries, especially in the heart. Arteriosclerosis (hardening of the arteries) is the greatest of all killers affecting heart, brain, ear, and kidneys. The human heart is the critical organ of your oxygen-burning machine. It dilates and contracts about a thousand times a day and approximately forty million times a year. The only rest that your heart muscles ever get is the fraction of a second between beats. The arteries must be flexible to do this enormous job! Don't block and harden them with inorganic minerals and toxic crystals from the wrong kind of food.

A man is as old as his arteries.
— Virchow

This catalytic agent — responsible for concrete — when taken into the body chemistry and following the process of natural metabolism through the years, becomes the principal mischief maker responsible for what is called "hardening of the arteries". Doctors call this degenerated condition of the arteries, "arteriosclerosis", which most people believe to be a natural condition that comes with the passing of the years. This is "herd" thinking — or rather, non-thinking. Yet few people question this age-old superstition, accepting it as fact that they must face arteriosclerosis and senility in the later years of their lives.

The very finest doctors will state that there is no known cure when hardening of the arteries takes place. Techniques have been developed to put plastic arteries in place of the larger arteries and veins of the heart and the neck, such as the carotid artery and the jugular vein. Also, there is very expensive surgery for cleaning out the inorganic deposits of some of the larger arteries of the body. But when you consider the entire pipe system of the human body, cleaning out a small amount could not mean a great deal. Miles of arteries, veins and capillaries would have to be cleansed of their inorganic crust to be effective.

BRAINS TURNED TO STONE

The greatest damage done by inorganic minerals — plus waxy cholesterol and salt (sodium chloride) — is to the small arteries and other blood vessels of the brain. Although modern science has developed some means to cope with the deterioration and terrible abuse of the kidneys, liver, heart and other important vital organs of the body, it can definitely be said that **no technique on earth can regain the life of a human brain that is turning into stone.**

What are premature ageing and senility but the brain turning into stone? Go to the large convalescent and rest homes and see with your own eyes the people who can no longer think or reason for themselves. Many of them cannot recognize their own children and relatives. All power of thinking is gone. They have no control over their eliminative organs. Adult diapers are put on them. Many of them have to be hand-fed. All the reflex power of the brain is gone. Their eyes stare into empty space. They are really dead people just waiting to be buried. These people are a pitiful sight — the most forsaken people on earth.

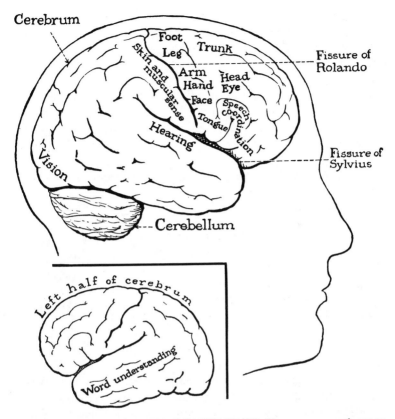

LOCALIZATION OF FUNCTIONS IN THE BRAIN. No power on earth can re-store the life of a brain turning into stone. Don't let your brain turn to stone. Drink vegetable and fruit juices and steam processed distilled water only.

And this end follows the exact pattern of the average person. Many humans are saved from this tragedy only because they die long before the body chemistry has had time to build their brains into stone. Hardening of the arteries and calcification of the blood vessels starts on the day you are born, because from birth we start taking inorganic chemicals into our bodies.

"Living under conditions of modern life, it is important to bear in mind that the preparation and refinement of food products either entirely eliminates or in part destroys the vital elements in the original material."

— U.S. Dept. of Agriculture

17

MY EARLY EXPERIENCES WITH HARD WATER

Take my own life, for instance. I was born on a farm in Vir-

ginia, along the Potomac River. We got all our drinking water from a well — crystalline, fresh, sparkling water. But it was very hard water, containing in suspension or solution — calcium carbonate and other inorganic minerals from limestone.

When we boiled this water, incrustations of these inorganic minerals formed in large slabs inside the kettles, and in time produced holes in the bottoms. Kettle after kettle had to be thrown away and replaced by another, with the same thing happening to the new one in time.

This hard water made dishwashing, laundering and cleaning difficult. The soap used for these purposes simply would not make suds.

Pure limestone
water.

But the greatest damage done by this hard water was to the humans who drank it.

My grandfather was a man in his mid-sixties. He was a big, strong six-footer, about 200 pounds of solid muscle. He was an expert horseman, a finished hunter and a hardworking farmer.

I can remember when he had his first major stroke. There was a large family of Braggs, and we were all seated

18

at the dinner table. Suddenly there was a crash of dishes, and my grandfather slumped over the table. When the country doctor arrived, he stated sadly that grandfather had lost all control of his left side due to brain damage.

From now on this poor man needed constant attention. With a completely paralyzed left side, he could not walk without the aid of someone to steady him. He had absolutely no control of his

SEVEN TYPES OF JOINTS

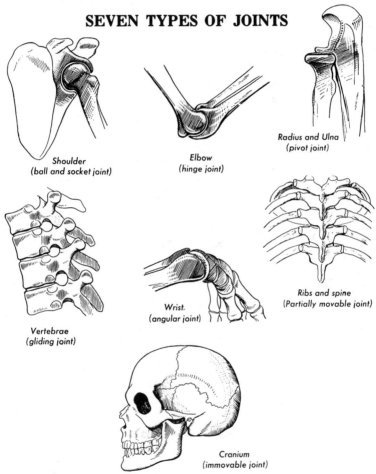

Shoulder
(ball and socket joint)

Elbow
(hinge joint)

Radius and Ulna
(pivot joint)

Vertebrae
(gliding joint)

Wrist.
(angular joint)

Ribs and spine
(Partially movable joint)

Cranium
(immovable joint)

These are the seven types of joints in your body that have movement. Between each of these moveable joints there is a clear amber fluid called synovial fluid which acts as a lubricant to keep the joints moving easily. When inorganic minerals from drinking water and toxic acid crystals replace this synovial fluid we have stiffness, pain and misery.

19

eliminative system. This helpless, sick man went into rages of anger. There was great difficulty getting food into his body because he had lost the ability to chew it. Only very soft, bland food could be fed him.

This fine man we knew and loved was, as far as real living was concerned, dead. You have no idea what a great burden he was on my parents and family. The poor, helpless man dragged on this way for three years; then the second and final stroke came and he was actually dead.

His body was sent to the Johns Hopkins Hospital in Baltimore, where the doctors who performed the autopsy stated that his arteries were like stone. My grandfather was born and reared on that farm and drank that hard water all the days of his life.

I was just a little boy when my father explained to me the outcome of the autopsy. I asked my father how his arteries could turn into stone. He could give no satisfactory answer to my question. And that very day I resolved to find out why my grandfather's arteries hardened into stone. I read medical books loaned to me by my Uncle William, who was our family doctor. I beseiged my doctor uncle with hundreds of questions as to why human arteries become hard.

It was to be many years before my questions were answered. In the meantime, I witnessed what the hard water was doing to my family and our relatives and friends.

Many fine black people worked on our farm. We all got along together as one big family. One of the black women who worked in our home was named Bessie Louise. She was just like a member of the family, and we all loved her dearly. Poor Bessie developed arthritis in her hands, wrists, elbows, hips, knees and ankles. How that poor woman suffered day after day with tormenting pain! Sometimes the pain would be so great that she would burst into tears.

Again I asked my doctor uncle what caused the arthritis. I wanted to know if there was a cure for this tormenting condition. He answered me honestly.

"Paul," he said, "we do not know the cause of crippling, painful arthritis, and we have no cure for it. The only thing I can do for Bessie is to give her strong drug painkillers to relieve her of her great suffering."

LOCATIONS IN THE BODY WHERE MISERY HITS THE HARDEST.

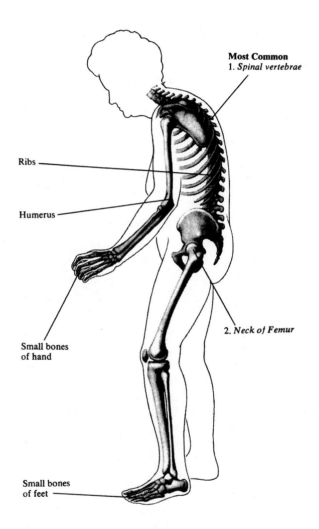

Most Common
1. *Spinal vertebrae*

Ribs

Humerus

Small bones
of hand

2. *Neck of Femur*

Small bones
of feet

In time, poor Bessie was confined to her bed of pain, and in a few years she was dead. She never reached 65 years of age. Her last days were days of intense pain and suffering.

My poor child brain suffered, too. What causes this horrible, crippling disease? I would ask myself in the late hours before going to sleep.

Here we were living on a big, fine farm, with an abundance of foods of all kinds. We had a good, comfortable home. It was a beautiful farm on a majestic river. But there was suffering among the adults. These pains were bulked into one word, and that was "misery". Each day I would hear my mother ask different people, "How is your misery today?" And the sufferers would give a doleful answer to my mother's question.

Again I would go to my kind, patient doctor uncle and put the question squarely to him. "Uncle," I would say, "why do so many people suffer from the misery?"

His answer was, "I wish I knew."

Then again I would say to myself, "Someday I will find out why people suffer from 'the misery'."

Millions suffer with pain!

T.B. IN MY TEENS

At the tender age of 12, I was put into a large military school in Virginia. My parents wanted me to prepare for West Point and make a military life my career. In my day and age, the parents did the thinking for the children. I did not want to be a soldier and I told my parents, but they told me they thought they knew what was best for me. And I obeyed.

At the military school I not only drank hard water, but I was fed a poor institutional diet — lots of starches, hot cakes, waffles, white rice, plenty of hots dogs, mashed and fried potatoes, overcooked meats, heavy desserts, doughnuts, sweet buns, chocolate cake, ice cream, pies, puddings, and desserts loaded with refined white sugar.

In exactly four years, at the age of 16, I was a victim of tuberculosis. Again and again I asked my doctor uncle why this had happened to me. Why? Why? Why? That good man could not answer my question.

I spent time in several T.B. sanitariums — and then fate stepped in.

I was in a hospital in New York, where four staff doctors examined me. I asked them pointblank, "Are you going to save me from this disease?"

And I received an honest answer. "No," they said, "we don't believe you're going to make it."

When they left the room, my little Swiss exchange nurse was angry. "These American doctors know nothing about T.B.," she declared. "I am glad I am returning to a sanitarium and a doctor who has cures."

I cried out, "Will you take me to that doctor? I want to live so I may help all sick people."

So the little Swiss nurse took me to her hospital in Switzerland where her great physician, Dr. August Rollier, gave me rebirth and a new life by using all natural healing methods — no drugs of any kind — just distilled water, good nutrition, sunshine, fresh air, deep breathing and exercise. In two years I was well and strong as a young stallion. Now I was ready to attain my life's ambition of helping others to help themselves!

THE SECRET OF RAIN WATER & SNOW WATER

Many of the practices of Dr. Rollier's sanitarium are now considered standard techniques in the treatment of all forms of tuberculosis. In many ways, my good doctor was 200 years ahead of his time.

One point on which he insisted was that no hard water was ever given a patient. Water is abundant in Switzerland, but Dr. Rollier gave us only rain water and melted snow water. He also was a great believer in the use of fresh vegetable and fruit juices.

And Dr. Rollier always told us the reasons for his treatments. He explained that practically all the water of Switzerland was heavy, or "hard" water, loaded with inorganic minerals which could bring nothing but harm to our bodies, because the body chemistry can assimilate only organic living foods and liquids.

I learned to love and revere Dr. Rollier because he gave logical answers to my questions. What a brilliant man! He brought healing to patients from all over the world who had been doomed to die, including myself. When I left the sanitarium, he cautioned me that I must drink only rain water, snow water, vegetable and fruit juices and distilled water.

THE ANSWER TO HEALTHFUL LIVING

Pondering Dr. Rollier's advice, I thought to myself, could it be possible that my grandfather's death from a stroke and Bessie Louise's death from crippling arthritis had a common basis. Was it due to drinking hard water and eating dead, devitalized foods? This question nagged at me. I had to find the answer to this question. I felt a great burden, a burden that could only be lifted when I found the truth; then the answer to disease and death would no longer be a mystery.

It was then and there that I resolved to be a biochemist, a nutritionist and a doctor who healed only with natural methods. After I left the sanitarium, I spent eight years in my search for knowledge which would equip me to help sufferers help themselves.

24

My work has been rewarding all these many years, and today I am more enthusiastic than ever about the powers of natural healing. That is why I have written this important book, which I think is long overdue, for now I have the answer.

MY FIRST TWO CASES

As noted, those two years at Dr. Rollier's sanitarium in Leysen, Switzerland gave me rebirth. Completely cured of my T.B., I was in excellent physical condition. The Alpine sunshine, the rain and snow water to drink, the pure, clean air of the Alps and the natrual diet had given me an entirely new body. Every cell in my body vibrated with vigorous health. Now I was ready to study biochemistry and other related health subjects to prepare for my life's work.

Deciding to live and study in London, I found a small apartment not far from the famous Regent Park. In my opinion, this is one of the most beautiful parks in the world. Here I could take my early morning runs and hikes and play tennis. In my apartment I could prepare my live food meals and live juices, and make my own distilled water.

The owner of the apartment building lived on the first floor. He and his wife were typical prematurely old people. They ate the regular English diet, which contained plenty of refined white flour (bread and its other devitalized products), large amounts of jams and jellies, gallons of tea with refined sugar and milk. Their vegetables and meats were all overcooked. To top it all, they drank London tap water, which was heavily chlorinated and chemicalized with calcium carbonate and other inorganic minerals.

When I came to inspect the fifth floor apartment — a "walk-up", no "lift" or elevator — the owner, Mr. Wilson, gave me the key and told me his joints were so stiff that he could not walk up the five flights. So I went up alone and found the apartment to be exactly what I wanted. Among other things to my liking, it was unheated. There were small, built-in grates; if I wanted heat I would have to order coal and have it delivered in bags.

I settled down comfortably in my fifth floor London apartment and started my biochemistry course. My landlords, the Wilsons, were very friendly, and from time to time I would drop into their apartment for a visit. Both of these nice English people had

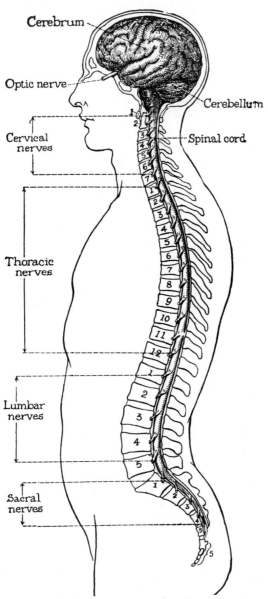

Cerebrum

Optic nerve

Cerebellum

Spinal cord

Cervical nerves

Thoracic nerves

Lumbar nerves

Sacral nerves

GENERAL VIEW OF THE CENTRAL NERVOUS SYSTEM AND SPINE
It is in the cushions between the bones of the spine that the inorganic minerals
from water may deposit themselves and cause back aches, slipped discs and
many other problems of the spine. Nerve force to the vital organs may be greatly
lessened, bringing on many painful miseries throughout the entire body.

26

many physical troubles. Mr. Wilson suffered greatly from pains in all his moveable joints plus a low back pain and some form of bladder disease. Mrs. Wilson was not much better off. She was fifty pounds overweight and huffed and puffed with every move she made. She suffered from a kidney disease. During my visits, a good part of our conversation centered around their ailments.

By this time the cruel London winter had set in. Outside it was damp and cold. But each day before dawn I would put on my heavy sweat clothes and take a long run in Regent Park, returning to the apartment glowing with health and well being. I never had as much as a sniffle all winter, but the Wilsons were plagued with one cold after another. They had large amounts of mucus and felt terrible.

One Saturday when I stopped by their apartment after my morning run, I could see that Mr. Wilson was desperately ill. He was running a high fever, and his nose was so completely stuffed up that he had to breath through his mouth. I went into his bedroom, which was over-heated and had very little oxygen.

The poor man looked at me and said, ''For God's sake, help me! I feel so sick.''

''Mr. Wilson,'' I told him confidently, ''if you will follow the natural system of healing that I will outline for you, you can get well.''

I knew I could help him — but I wondered before making my offer? — if he was strong of mind and wanted health strongly enough!!!

''I will follow your instructions to the letter'' he stated in desperation like a drowning man grasping a helping hand.

''Good! Today you will start on a ten-day fast.'' I picked up the bottles on the bedside table. ''All this medication will go down the drain.''

I brought him some of my distilled water, purchased lemons and honey for him, and the fast started. It was not easy for this man to fast. He was so full of toxic poisons, so full of sticky mucus in his head, throat and lungs, that he had a great deal of discomfort in getting rid of it. But he was an Englishman with plenty of stick-to-it-ive-ness. He passed from his body lots of toxic wastes. And at the end of the ten-day fast, he felt better than he had for many years.

Then I put him on a natural live food diet and gave him fresh fruit and vegetable juices and distilled water. Within three weeks after his fast, he climbed the five flights of stairs to my apartment — something he had not done in seven years.

His wife became enthusiastic about the natural way of living and started to shed the fat that covered her body.

In six months you would never know the Wilsons were the same people. Mr. Wilson bounded up to my apartment several times a day, two steps at a time. Mrs. Wilson looked trim and slim and had to have all her clothes made smaller. Their married daughter, who was living in Canada, came for a visit and could not believe what she saw. The Wilsons' troubles were gone. They were enjoying life to its fullest.

I was happy, too.

The Wilsons thanked me and told me they felt as if they had been "reborn". By following the Natural laws of health, they found vigorous health.

These were my first two cases. The results gave me great confidence. And confidence grew more and more as I studied the teachings of the great healers of the world — to even Hippocrates (Father of Medicine) wise words he gave to the world to use — **"Let food be your medicine and medicine be your food."**

NATURE'S WAY

Now as I think of the thousands of people who have come to me for advice, and the many people who have been reborn by living Nature's way, it gives me deep satisfaction to know that I have been able to help people really live in health again.

This is Nature's own way of life. There is a big difference between feeling well enough to carry on one's daily activities with no sensation of anything wrong, and that more exaggerated state of health which fills one with enthusiasm for life and its challenges. An adequate amount of that important commodity, vibrant health, supplies sufficient energy for life to go on

28

serenely, but it takes more to give one a sense of exuberance.

People who live by Nature's way enjoy an exalted feeling of well being that is not euphoric (abnormally happy and buoyant), but is a natural **joie de vivre,** or joy of living.

The Wilsons found this joy of living when they gave up chemicalized, inorganically mineralized water and went on a natural diet. They learned through their own experience that the body is a self-healing and self-repairing instrument. Mr. Wilson found out that the stiffness in all moveable joints was not due to the number of years he had lived. He learned that time is not toxic. His stiffness was brought on by a combination of toxic acid crystals from his unbalanced, acid diet and drinking water saturated with inorganic minerals and chemicals. Fasting helped to dissolve these incrustations which had been deposited in his joints. Natural diet and distilled water continued the healing process and helped prevent recurrence of his former ailments. The same thing happened with Mrs. Wilson's problems of overweight and kidney trouble. By making a complete change from their old, incorrect life pattern to a natural way of living, they were able to enjoy the full potential of healthful energy — the true joy of living.

THE STONES WITHIN US

The more I learned about biochemistry (life chemistry), the more I realized why so many people were prematurely old and suffered pain all over their bodies. On my visits to the large London hospitals, I learned more about stones forming within the human body.

Why do stones form in the body, and what does this mean in regard to human health?

The most common places to find such stones are in the gall bladder, the kidneys, the passageways between kidneys and bladder (known as the ureters) and in the bladder itself. Another organ in which stones are sometimes seen by X-ray is the pancreas, the glandular organ which lies behind the stomach and has both an internal and an external secretion.

Stone formation anywhere in the body has always been regarded as a diseased condition.

In my opinion, all these stones are formed by the unbalanced, acid, toxic-producing diet that most humans eat, plus the chemicalized, inorganically mineralized water they drink, plus the heavy concentrations of salt most people use, plus the tremendous amount of waxy cholesterol (saturated fats) eaten by the average person. Unbalanced diets form toxic poisons which the body cannot eliminate, so these are formed into stones by the body's chemistry. Practically all drinking water contains the inorganic mineral calcium carbonate. This and other inorganic minerals play their part in forming stones within the body's vital organs.

GALL STONES — SILENT AND NOISY

"Silent" gall stones are those which remain quietly in the gall bladder and thus do not produce the acute abdominal pain which is known as gall stone colic. However, these silent stones may at any time — and perhaps very inopportunely — become raucously "noisy".

Noisy gall stones may involve not only the gall bladder itself but also the common duct, a vital structure which carries secretions from both the gall bladder and the liver into the intestine. This often happens when the gall bladder contracts and attempts to expel a gall stone. If the stone gets stuck on the way out, there is acute pain and frequently inflammation of the gall bladder and the common duct.

If the stone blocks the common duct, the liver cannot send its bile into the intestine where it is essential for proper digestion of food. The liver itself is then in trouble. The result is what is known as obstructive jaundice, the

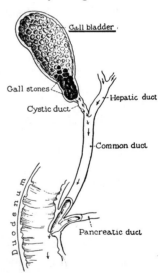

GALL-STONES IN THE GALL-BLADDER
Gall-stones may be caused by the drinking of water saturated with inorganic minerals and the toxic crystals from an unbalanced, unnatural diet. Over-eating of saturated and hardened fats may cause disturbance in the Gall-Bladder.

yellow of the bile showing in the skin and in the whites of the eyes.

The color of the skin also betrays the presence of silent gall stones. This was the case with the great Hollywood film actor, Tyrone Power, who came to me for counselling. He was a fine specimen of a man with tremendous ability, but I could tell from the color of his skin and his eyes that he was suffering with silent gall stones. I pleaded with him to change his habits of life and follow Nature's way, but I could not communicate my health message to him. The poor fellow died at a very young age. Had he let me give him a program of liver detoxication — and had given up his unbalanced diet, salt and ordinary drinking water — that talented and handsome man would be living today.

I have had many people under my nutritional supervision who had gall stones. And the Vital Force of their bodies was so increased that the gall stones were safely squeezed out through the common duct into the small intestine and eliminated.

KIDNEY STONES

In my opinion, the cause of many kidney stones is heavy, hard, chemicalized water, saturated with calcium carbonate and other inorganic minerals.

Beneath my home in the California desert, there is a subterranean river several hundred feet below the surface of the earth. When wells are sunk into this river, the water comes out at a heat of 175 degrees. It is heavily saturated with calcium carbonate and its chemical affinities such as magnesium carbonate. This water is not permitted to flow through cast iron or steel pipes, because the incrustations of the inorganic minerals will soon block the core. Only copper pipes are used for plumbing.

People come from all over the world to bathe in the waters in this village of spas. The hot water does have a curative value. One thing for sure, it brings relief to those suffering from arthritis and rheumatism. Most of the hot water pools are kept at a temperature of 104 to 108 degrees. Body heat is 98.6 degrees. When you submerge the body in water hotter than body temperature, you start an artificial fever, and many toxic poisons are eliminated through the 96 million pores of the skin. We all know that a good sweat is refreshing to the body. We always feel lighter.

The very sad part about coming to this hot water resort, however, is that people are also advised to drink this heavily saturated inorganic mineral water. And let me tell you, the concentrations of these inorganic minerals are extremely high. If you put five gallons of this mineral water in a pan and let it evaporate in the sun, you will have a mass of inorganic minerals.

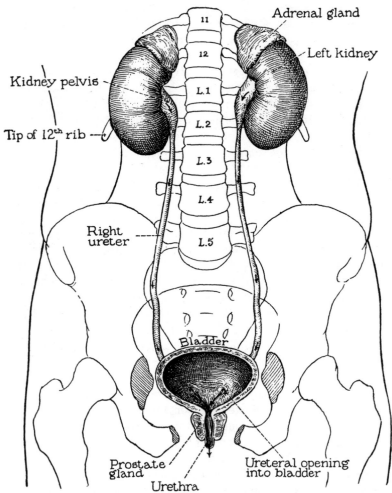

THE URINARY SYSTEM — KIDNEYS, URETERS, AND BLADDER. The Adrenal bodies are shown on top of the kidneys. It is in the urinary system that inorganic minerals and toxic acid crystals may cause kidney and bladder stones. The urinary system must be kept free of these deposits to remain in the elastic condition characteristic of youth.

DON'T DRINK INORGANIC MINERALIZED WATER

Several years ago a gentleman from New York came to this hot water resort to take the baths. The uninformed owners of the spa told this man also to drink the mineral water as it would be good for him. I advised him strongly to bathe only — not to drink the water. But he did not heed my advice. During the six months he took the baths, he also drank this water of death. One night the people in the hotel heard him scream out in agonizing pain. When they reached him, he was dead. The autopsy showed that he was killed by a large kidney stone puncturing a large artery.

Millions Die Before Their Normal Time!

Thousands upon thousands of people all over the world have kidney stones of various shapes and sizes. Sometimes the stones get so troublesome that one kidney must be removed by surgery.

I have visited hot and cold water spas all over this country and many other parts of the world. The operators of these spas tell people that by drinking and bathing in these waters, this and that disease will be cured. This I do not believe. Relief of pain by mineral water bathing — yes. Detoxification of body wastes by mineral water bathing — yes. But drinking this heavy inorganic mineral water only brings on serious trouble.

My sincere and honest advice to you is: **Don't drink inorganic mineralized water!**

Just keep in mind always that you cannot assimilate inorganic minerals. You can only assimilate organic minerals which come from that which is living or has lived.

DOLOMITE TABLETS ARE INORGANIC

During the past few years Dolomite Tablets have come on the market. Let's look at an advertisement by one of the producers of these tablets:

"After many months of research and exhaustive tests, this company has developed the original natural DOLOMITE TABLETS extracted from mineral-rich earth . . . This super-fine, extra-smooth powder contains high concentrations of calcium and magnesium, and is excitingly different from other limestone products. Our DOLOMITE TABLETS are formulated from a specific form of LIMESTONE, known as DOLOMITE. This is exceptionally high in magnesium. It contains a high percentage of magnesium in combination with calcium. This product is mined from the mountains and earth."

In no part of this advertisement does the producer of these Dolomite Tablets say that they are made from inorganic calcium and magnesium. He does state that they are formulated from a specific form of limestone.

These minerals are inorganic, and regardless of how they have been powdered and baked, they cannot become part of the organic chemistry of the body. These Dolomite Tablets are the same formula that was in the cattle feed which was purchased by my father many, many years ago on our Virginia farm. As I said previously, this inorganic mineral cattle feed could not be used by our cattle — they refused to eat it; and it was finally taken off the market.

Similarly, these Dolomite Tablets cannot be used by the human body chemistry. Only organic foods can be used by the body. Beware of any pill that is extracted from "mineral-rich earth"! To buy it would not only be wasting your money but also asking for physical troubles.

When pure rules of business and conduct are observed, then there is true religion. Walk . . . path of duty, do good to your brethren, and work no evil towards them.

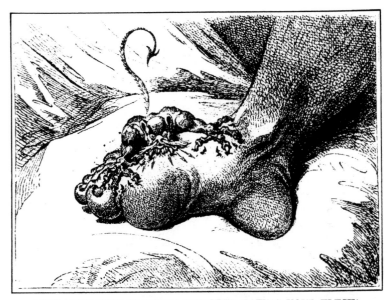

GOUT: THE PAIN IS LIKE A MONSTER EATING YOUR FLESH!
Cartoon from an old English drawing.

WHAT IS GOUT?

People are sometimes disturbed when a doctor makes a diagnosis of gout to explain an aching joint, especially in the big toes. Perhaps they remember the old pictures of the British Lord with one leg wrapped up and propped on a chair in front of him and, of course, with a look of great pain on his face. They also remember that he arrived at this unhappy state via living high on the hog — a diet high in animal flesh, heavy in eggs, milk and cheeses, rich sauces and gravies made from meat, and all washed down with chemicalized and inorganically mineralized water.

For more than 65 years I have seen the high protein diet come and go. The sponsors of these diets rationalize that we are made of protein, therefore we must eat large amounts each day to build our bodies and our strength.

Meat protein is heavily saturated with a powerful toxic material called uric acid. Gout is due to a distrubance in the production, destruction and excretion of uric acid.

After various chemical operations have gone on in the body to break down the proteins that are found in all living cells, this substance — uric acid — is a final or end product. A certain amount of it is found normally in the blood, up to about five or six milligrams per 100 milligrams of blood serum. When this amount is exceeded, a tendency to gout — if not the disease itself — is evident.

How do you know if you have it? If you have severe pain in one of your joints, most often the great toe joint, your doctor will suspect that this is a special form of joint inflammation (arthritis), known as gouty arthritis or simply gout. As distinguished from chronic arthritis, there is no residual pain or tenderness between severe attacks.

If the disease is unchecked, the periods between attacks become shorter and the joint gradually becomes deformed. Toxic crystals formed from uric acid, as well as from inorganic minerals in heavy drinking water, are deposited in joints or bursa, then inflammation and destruction result. These same deposits are also found in cartilage surrounding bone anywhere in the body and are called chalk stones or tophi, a characteristic finding in gout.

The kidneys are often involved in these disturbances. Structures of the kidneys called tubules may be blocked by these crystals from uric acid and inorganic minerals. These toxic crystals are often re-absorbed into the body from the kidney tubules, aggravating the trouble.

BALL AND SOCKET

It is within the shoulder, elbow, wrist and hand joints that inorganic minerals from salt, heavy water and toxic acid crystals from wrong food may form and cripple one or more of these joints, causing pain and restricted movement.

HINGE

SLIDING

In fact, the most serious complication of gout is kidney damage.

FOUR TYPES OF JOINTS

Ball-and-socket joints at hips and shoulders permit freest movement of all body joints. Hips and shoulders are examples

The vertebrae are saddle joints, moving forward, backward, and sidewise. One vertebra moves only slightly on the next, but the whole column is fairly flexible

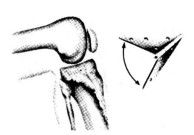  

Hinge joints are like the hinges you know— permitting backward and forward movement only, like the hinges of a door. Your knees and your fingers are hinge joints.

Pivot joints permit the bones to rotate at the joint like a key turning in a lock. The elbow is a combination of pivot joint and hinge joint. Thanks to this joint one bone of the forearm can rotate about the other

What can we do about this painful and distressing condition? In these pages it is not possible for me to offer cures — this book is not written to cure but rather to tell you what kind of health program to follow to help nature clear up this painful condition.

If this misery hits you, the first thing to do is to fast. Fast at least one week on distilled water (hot or cold) flavored with fresh lemon juice and ¼ to ½ teaspoon of honey if desired. Drinking large amounts of pure, steam-produced distilled water helps the kidneys to cleanse themselves and often prevents the formation of kidney stones (which also come from uric acid and inorganic drinking water).

After the distilled water fast, your diet should exclude products high in purine. This is a chemical which is called the parent of the uric acid substances. The person suffering from gout should eat no kidneys, liver, sweetbreads, sardines, anchovies

37

or meat extracts. Even meats, fish, pork, pork products, fowl, peas, beans, lentils, cheese, eggs, milk and milk products should be discarded.

FRUITS AND VEGETABLES — NATURE'S FINEST

The most protective foods are raw fruits and vegetables. About 60 percent of the diet should be raw fruits and vegetables and their fresh juices. Properly cooked vegetables, sunflower and sesame seeds should constitute the protein foods. Even whole grain breads should be eliminated for at least a year. A weekly 24-hours distilled water fast should be faithfully taken.

ARTHRITIS AND RHEUMATISM

There is a good bit of confusion in the use of the words rheumatism and arthritis. Nowadays the word rheumatism is used loosely to mean pain and discomfort in and around the joints. In stricter usage, the rheumatic miseries include not only those of bone and cartilage, but also of tendons and tissue surrounding bones, or connective tissue. We also use the word bursitis when the inflammation is confined to the bursa, a sac containing fluid to prevent friction between joint and tendon.

In connection with the general term, rheumatism, it is of interest to note that one in twenty persons in the United States has rheumatic complaints, making this one of our common physical miseries. About half of these people have actual arthritis, and of these, one-tenth are disabled to some degree.

All this indicates that arthritis, which strictly means inflammation of the joints, is a justly dreaded misery. It has been estimated that there are more than fifty varieties of this disease. The kind most feared is known as rheumatoid arthritis. All ages may be affected. Even very young children suffer from this deforming misery.

38

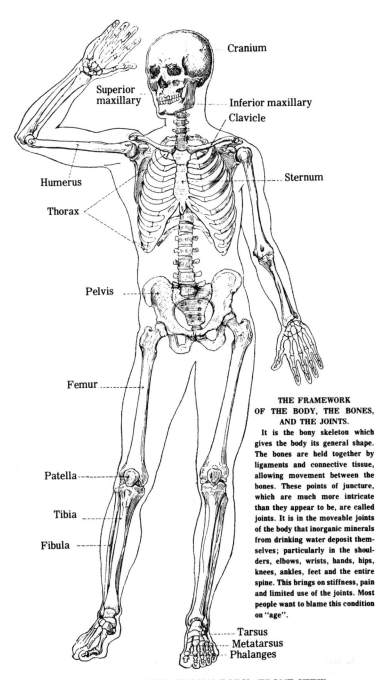

Cranium

Superior
maxillary

Inferior maxillary

Clavicle

Humerus

Thorax

Sternum

Pelvis

Femur

THE FRAMEWORK
OF THE BODY, THE BONES,
AND THE JOINTS.

It is the bony skeleton which gives the body its general shape. The bones are held together by ligaments and connective tissue, allowing movement between the bones. These points of juncture, which are much more intricate than they appear to be, are called joints. It is in the moveable joints of the body that inorganic minerals from drinking water deposit themselves; particularly in the shoulders, elbows, wrists, hands, hips, knees, ankles, feet and the entire spine. This brings on stiffness, pain and limited use of the joints. Most people want to blame this condition on "age".

Patella

Tibia

Fibula

Tarsus
Metatarsus
Phalanges

THE BONES OF THE HUMAN BODY. FRONT VIEW.

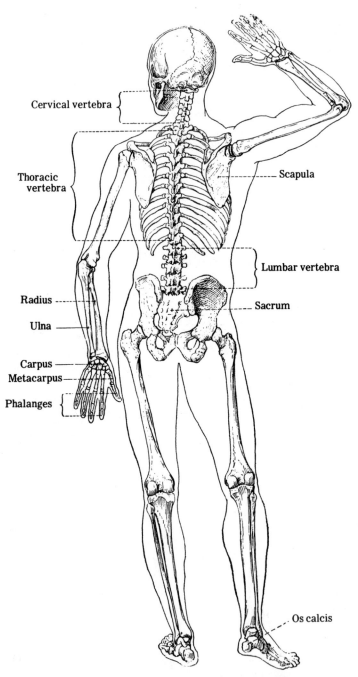

Cervical vertebra

Thoracic vertebra

Scapula

Lumbar vertebra

Radius

Sacrum

Ulna

Carpus

Metacarpus

Phalanges

Os calcis

THE BONES OF THE HUMAN BODY. BACK VIEW.

Rheumatoid arthritis may affect different parts of the body, but the joints are the chief targets. There is inflammation in one or many joints, its onset characterized by redness, heat and swelling. When a joint is swollen and painful, it is difficult to use it, and it therefore becomes less flexible from lack of use as well as from the misery itself. The muscles also grow smaller without being used, and the victim may seem to have large and very sore joints with thin arms and legs.

There is no known cure for rheumatoid arthritis. I have no cure to offer you. Again, all I can offer you is a Health Plan For Living. Only the basic biological functions of the body can help correct this misery.

HEALTH HINTS FOR RELIEF AND BENEFITS

In order to prevent those important muscles of yours from shriveling and becoming useless, they must be exercised — but only the correct way. If you don't use them (your 600 muscles) — you lose them! Besides keeping the muscles from wasting away, the gentle exercises expertly prescribed for you are designed to preserve the moveability of the joint. Even if it pains you to exercise the affected areas — you must try — and gradually work the muscles and joints loose and free of the toxic (cement-like) crystals so they can be dissolved and be thrown off by the elimination system.

Great relief from pain and swelling may be obtained through heat. Heat relieves the spasm of the muscles and thus improves the blood flow to both muscle and joint. Usually it is best to apply heat (hot bath or heating pad) for a little while before exercise. This helps relax the area and loosens it up — so you will have an easier time of exercising. At the great hot mineral water spas near my home in California, I see many helpless sufferers from rheumatoid arthritis getting blessed relief from the hot mineral water. If you cannot go to a hot spring for relief, you can take a hot epsom salts bath (one cupful of epsom salts to a tub of very hot water).

Bed boards are helpful to prevent the spine from taking on a curve from a soft sagging mattress. Personally, I sleep on a no-spring bed — just a thin mattress over a wood platform. This is a great way to keep the spine supple and strong.

POSTURE SILHOUETTES

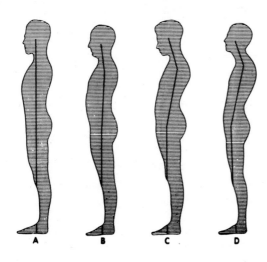

A B C D

(A) Good: head, trunk, and thigh in straight line; chest high and forward; abdomen flat; back curves normal. (B) Fair: head forward; abdomen prominent; exaggerated curve in upper back; slightly hollow back. (C) Poor: relaxed (fatigue) posture; head forward; abdomen relaxed; shoulder blades prominent; hollow back. (D) Very poor: head forward badly; very exaggerated curve in upper back; abdomen relaxed; chest flat-sloping; hollow back.

Postural exercises to prevent a curved back and a stooped neck are excellent. In fact, this type of exercise should be done by everyone, even those without any sign of arthritis, to keep the erectness of youth as long as possible. Walk tall, stand tall and sit tall — make your muscles work to keep you stretched up as tall as possible at all times! All great men of history had good posture — and you can acquire it also — with practice — start now! (Practice makes perfect!)

Again, let me state emphatically that, in my opinion, the misery of arthritis is caused by hard water saturated with inorganic minerals and an unbalanced diet, forming acid crystals in the moveable joints, plus inactivity of the body in general. It is a combination of unnatural living habits. Every effect must have a cause! There is a reason why things happen in the body. Failure to live Nature's way is the cause of most human miseries.

"To preserve health is a moral and religious duty, for health is the basis for all social virtues. We can no longer be useful when not well."
— Dr. Samuel Johnson, Father of Dictionaries

DON'T LET YOUR BRAIN TURN TO STONE

My neighbor James K. is 65 years old. You will note that I said 65 years **old** — not 65 years **young**. Jim will be forced to retire from his position as an executive in a large company in a few months. This is a strict company rule.

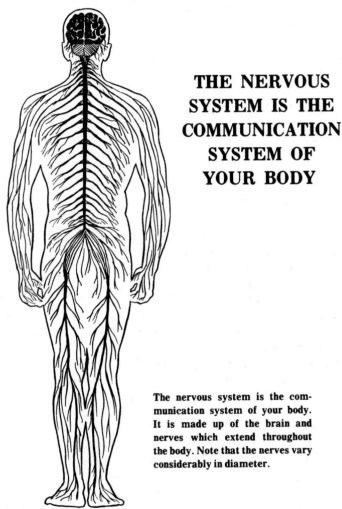

THE NERVOUS SYSTEM IS THE COMMUNICATION SYSTEM OF YOUR BODY

The nervous system is the communication system of your body. It is made up of the brain and nerves which extend throughout the body. Note that the nerves vary considerably in diameter.

Why do so many large corporations require all employees to retire at age 65? The reason is that by 65, most people have hardening of the arteries of the brain. The brain has lost much of its blood supply and is not getting the life-giving oxygen that makes it keen, sharp, creative, wide-awake and positive.

Remember that many of the pipes supplying the brain are as small as the hairs of the head. Years of drinking chemicalized, inorganic mineral water and years of eating a highly unbalanced diet heavy in salt, incrustations and toxic acid crystals have clogged and hardened the arteries, veins and capillaries that must supply the brain with oxygenated blood.

There is a definite link between physical vigor and mental vigor. It all comes down to the fact that we must have a sound mind in a sound body.

People actually build rock formations in the blood vessels supplying the brain, just the same as the great rock formations are made in limestone caverns drop by drop. You can see how these great columns of stalactites and stalagmites are formed by inorganic mineral water, one drop at a time. The brain does not turn into stone in a few years, but year after year of drinking inorganic mineral water and eating toxic foods builds the rock formation in the human brain.

DOOMED TO THE HUMAN SCRAP PILE

Big corporations will not accept an application from a man or woman who is over 50 years of age. They know from actual ex-experience that there is considerable deterioration of the brain in people of 50 and over.

It all boils down to simple physics. You have small pipes leading to the brain. The way the average person eats and the water he drinks bring on degenerative changes in these pipes and in the brain itself. The longer the average person lives, the more degenerative changes take place in the brain.

Many senior citizens will admit that their brain power is slipping. They will tell you how poor their memory is, how they cannot recall names and events. A brain turning to stone does not have the capacity to be wide-awake and sharp. As this condition gets worse, we call it senility. In time, the brain solidifies to a point where it remembers nothing. This is called living in deep senility. Or is it living death?

"Men do not die, they KILL themselves."
—**Seneca, Roman Philosopher**

HOW THE BRAIN FUNCTIONS

Did you ever stop to think what makes you think?

Inside the protective covering of the bony skull is a mass of what we call "gray matter". Gray matter is tissue composed of millions of nerve cells "woven" together so that what we see, hear, smell, taste and touch gives us an awareness of our status on this earth.

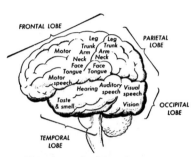

These are the control areas of the brain. Each thing you do, such as seeing, hearing, speaking, or moving, is controlled by a certain part of your brain.

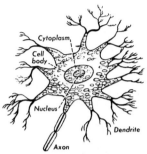

The entire nervous system is made up of individual cells called neurons. Every neuron, as shown in this drawing, has three main parts: the cell body, the dendrite and axon.

As the average person's brain slowly turns to stone, much of this birthright of keen awareness is lost. The sight starts to go. Cataracts develop, actually a stone formation over the eyes. The hearing becomes impaired because the arteries leading to the ears become corroded with incrustations of inorganic materials. These are diseases of degeneration. This is what we blame on the passing years — not on how we have lived.

This gray matter we call the brain must have oxygenated blood, or it degenerates. All living cells of the body must have oxygen in large quantities to survive. Senility is actually oxygen starvation of the brain.

With that "gray matter" we also think, know, remember, judge and believe. It was named gray matter because it is in part pinkish-gray in color although the brain also has a white part. Our behavior and emotions are controlled by that mass of tissue. We now know that secretions of the endocrine glands also enter into the stream of communications which affect the brain cells.

This structure, the brain, is an unbelievably complex electro-chemical organ, as variable in man as are fingerprints. It is the miracle of life — a miracle which gives us joy and sorrow, philosophy, understanding, reasoning power, will power and the ability to have feelings. Philosophers have called it "man's unconquerable mind". As we see what has been developed by it from one century to another, we can almost call that adjective "unconquerable" a factual one.

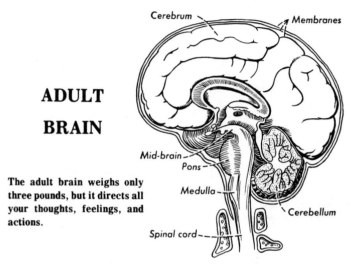

Cerebrum

Membranes

ADULT

BRAIN

The adult brain weighs only three pounds, but it directs all your thoughts, feelings, and actions.

Mid-brain

Pons

Medulla

Cerebellum

Spinal cord

To have the unconquerable mind, we must constantly provide it with a free flow of rich, red blood carrying the life-giving oxygen. That is the reason the supply pipes leading to the brain must not be blocked with inorganic minerals. If you wish to regain and maintain a strong brain use only steam-produced distilled water, and fresh fruit and vegetable juices, as your drinks. Keep far away from city waters, alcohol, tea, coffee, cola and soft drinks.

THE BRAIN NEEDS EXTRA FINE NUTRITION

The brain must be adequately nourished in order to function. No other part of the body fails more quickly from lack of good nutrition. And upon what does this marvelous structure feed?

It needs foods rich in enzymes. Raw fruits and raw vegetables and their fresh juices provide excellent nourishment. Soy beans, which are exceptionally rich in lecithin, should be eaten several times weekly. Lecithin (powdered, liquid, capsule, tablet or granule form) can also be purchased at your Health Food Store. Sunflower seeds, sesame seeds, pumpkin seeds are all healthful brain foods.

ORGANIC MINERALS ESSENTIAL TO LIFE

The brain needs phosphorus. Organic phosphorus is found in beans of all kinds such as pinto beans, garbanza beans, dried lima beans and lentils. Other sources are all 100% whole grains, brown rice, almonds, peanuts and walnuts. Lean meat, egg yolk and natural, unprocessed cheese contain organic phosphorus.

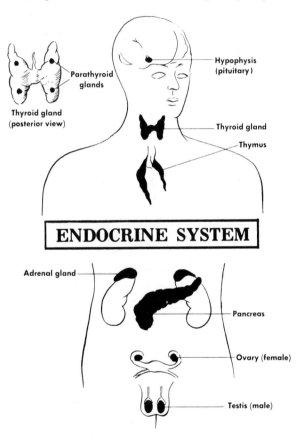

Parathyroid glands

Hypophysis (pituitary)

Thyroid gland (posterior view)

Thyroid gland

Thymus

ENDOCRINE SYSTEM

Adrenal gland

Pancreas

Ovary (female)

Testis (male)

All the organic minerals are needed to keep the body strong, youthful and healthy. They are essential factors in digestion and assimilation and important ingredients of the digestive juices, regulating the osmotic exchange between lymph and blood cells. In short, organic minerals are indispensable to the proper physiological functioning of all the glands of the body.

It is estimated that a normal man weighing 150 pounds is composed of the following:

90 lbs. oxygen
36 lbs. carbon
14 lbs. hydrogen
3 lbs. 8 oz. nitrogen
3 lbs. 12 oz. calcium
1 lb. 4 oz. phosporus
4 oz. of chloride
3½ oz. of sulphur
3 oz. of potassium
2½ oz. of sodium
2 oz. of fluorine
1½ oz. of magnesium
¼ oz. of silicon
1/6 oz. of iron

Trace Elements

There are also traces of these important elements: Manganese, aluminum, iodine, copper, lead, zinc, lithium, cobalt, helium, neon, etc.

MINERALS MAKE
THE MAN!

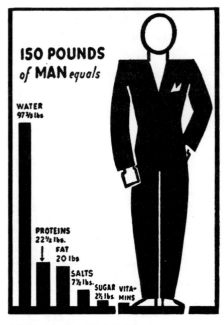

150 POUNDS *of* MAN *equals*

WATER
97⅓ lbs

PROTEINS
22½ lbs.
↓ FAT
20 lbs
SALTS
7½ lbs. SUGAR VITA-
2½ lbs. MINS

FORMULA FOR A HUMAN BEING

According to B. A. Howard — in his book The Proper Study of Mankind — the human body contains:

- Enough **water** to fill a 10-gallon barrel;
- Enough **fat** to make 7 bars of soap;
- Enough **carbon** for 9,000 lead pencils;
- **Phosporus** enough for 2,200 match-heads;
- **Iron:** just enough for one medium-size nail;
- **Calcium** (lime)) enough to whitewash a chicken coop;
- And microscopic amounts of such trace elements as cobalt, iodine, zinc, copper, molybdenum, titanium, beryllium, etc. Take these ingredients, combine them in the right proportions, in the right way, and the result, apparently, is a man.

48

Remember, these are **ALL ORGANIC** — not inorganic — chemicals and minerals. There is a sharp line of demarcation between the two. Although the chemical analysis is the same whether found in air, earth, plant or animal, it is only through

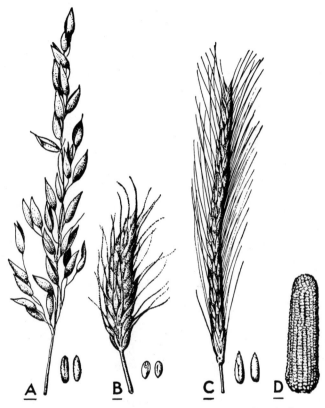

These foods are rich in organic minerals.

A, oats. B, wheat. C, rye. D, corn.

the life process of the plant that the constituents of air and soil become vitalized. It is this property of vitality alone which distinguishes, for example, the atom of iron in the red corpuscles of the blood from that of inorganic iron or preparations made from inorganic iron. You could suck on an iron nail for years and never extract any organic iron for building your blood. When you eat blackberries, you are getting organic iron that can be used by the blood. The arrangement of atoms that form a molecule of

49

the iron nail is the same as that of the organic iron in the blackberry. Only by the great natural force of photosynthesis does the living plant convert the inert inorganic minerals into the organic minerals which man can use for keeping himself alive.

Sometimes the minerals of the body are referred to as "mineral salts". This misleading terminology has given the public the erroneous idea that the term "salt" refers to common table salt, or inorganic chloride of sodium, which most people mistakenly consider an indispensable adjunct to almost all foods.

The fact cannot be too frequently emphasized that there is a vital change going on in all the minerals as they pass into the structure of the plant. On the other hand, chemical analysis or separation of the minerals means destruction of the living tissues. Of course, the chemist will find in the minerals of the "ash" the same properties that are found in the minerals of the soil. But the subtle, imponderable force — vital electricity — has escaped him. It cannot be isolated by the laboratory process of condensation or extraction. We must learn to recognize the mineral elements of the body as really being "organic" — integral parts of the living body and subject to the same vital changes, life and death, as the entire organism.

The organic calcium of the skeleton, the organic iron contained in the red corpuscles, the organic sodium and potassium found in the blood serum are organized, and as such have a certain duration of life during which they have vital functions to perform. Sooner or later the molecules will lose their electromagnetic tension, according to the degree of their physiological activity. In other words, they have served their purpose and must be supplanted by fresh minerals. That is the reason that 50% of your diet should be live, living raw fruits and vegetables. These are the great suppliers of the imponderable force — vital electricity.

THE ALKALINE OR BASE-FORMING MINERALS

These are the eliminators of toxic waste poisons, the real immunizers of the body.

The alkaline minerals, which are so important in the performance of the physiological functions of the body, are iron, sodium, calcium, magnesium, potassium and manganese. These are essential in the formation of the digestive juices and the secretions of the ductless glands, the hormones which probably regulate nearly all the vital processes of the body.

Iron is necessary for the formation of red blood corpuscles and is the oxygen carrier of the system.

Elimination of carbon dioxide depends largely upon organic sodium, which is the chief constituent of the blood and lymph.

Calcium, combined with magnesium, phosporus and silicon, makes up more than half the bony structure of the body and imparts textile strength to all tissues. It also serves as a neutralizer and eliminator of toxic poisonous acids.

Remember, whenever we refer to these minerals in body chemistry, we are speaking of **organic** minerals.

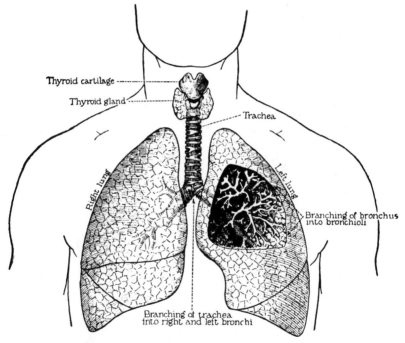

THE LOWER RESPIRATORY SYSTEM. This is where inorganic minerals and toxic acid crystals, that may cause serious trouble in the Respiratory System, deposit themselves shutting off the supply of life giving oxygen.

HEALTH in a human being, is the perfection of bodily organization, intellectual energy, and moral power.

— T. L. Nichols, M.D.

IRON, THE OXYGEN CARRIER OF THE BLOOD

Organic iron is indispensable for the formation of chlorophyll and hemoglobin.

On account of its great affinity for oxygen, iron plays an important part in the organic world and stands in very close relationship to the fundamental processes of change of matter, metabolism.

The plant or tree takes the inorganic iron from the soil and carries it to the leaves where it takes part in the formation of the chlorophyll granules, the green coloring matter of nature. The amount of organic iron and chlorophyll varies in the different parts of the plant. For instance, the green outer leaves of cabbage contain four times as much iron as the inner etiolated leaves.

HOW PLANTS DO THEIR WORK

In order to carry on its life processes, every organism is equipped with structures enabling it to use the materials in its environment for the satisfaction of its own needs. Animals, with their power of locomotion, are able to go in search of food. Plants, not having this power, must have some way of procuring their food from their immediate surroundings. In higher plants the structures particularly fitted for this purpose are the root, stem, and leaf. The **root,** besides anchoring the plant in the soil, takes in water and minerals. The **leaf,** being rich in chlorophyll, is able to carry on the process of photosynthesis by uniting the water taken in by the roots with the carbon dioxide of the atmosphere, thus producing simple sugar, an organic food, for the plant. The **stem** is an intermediate structure which conducts the water from the root to the leaf and holds the leaf in a position in which it can receive maximum sunlight. It also carries the newly manufactured sugar from the leaf to various places in the plant where it can be stored.

In the Bodies of Plants, Animals, and Man, Iron Serves the Following Distinct Purposes:

1. To produce the chlorophyll of the plant, principally contained in the green leaves, and the hemoglobin of the red corpuscles.

2. To enable the plant to take carbon dioxide and nitrogen from the air and to synthesize them into organic matter by means of chlorophyll and sunlight.

3. To assist in the processes of respiration in man and animals. It is the hemoglobin that carries the oxygen to all parts of the body, reaching every cell through the capillaries. Here the carbon of the ingested food, stored in the cells of the tissues, is oxidized and changed into carbonic acid. This in turn is combined with the alkaline elements of the blood and eliminated through the lungs.

4. To generate a magnetic blood current and an electromagnetic induction current in the nerve spirals which pass through the walls of the arteries and veins and help to build and nourish the tissues.

The total amount of iron in the human body is comparatively small, probably not exceeding 75 grains under normal conditions. Of this quantity about 50 grains are contained in the blood, the remainder being distributed in the marrow of the bones, the liver and principally in the spleen. Iron is the most active mineral in the system, and therefore needs to be renewed more frequently than the more stable elements of calcium and potassium in the bones and tissues.

The quantity of blood of a normal adult man of 160 pounds is about 12 pounds (7½% of body weight), containing approximately 50 grains of iron. With every pulse beat, nearly six ounces of blood are forced from the heart into the major artery, the aorta; and during every half minute, the entire quantity of blood passes from the heart into the lungs and from there into the arteries and capillaries through the body. Consequently, the 50 grains of iron pass through the heart and lungs 120 times per hour, or 2,880 times per day. Within 24 hours under normal conditions, the 50 grains of iron have to perform the same function as 2,880 x 50 grains, or more than 20 pounds. For that reason a daily supply of organic iron in our food is essential.

The best organic iron sources are greenleaf vegetables such as watercress, raw spinach (do not eat cooked spinach as too much oxalic acid is produced by heat), raw parsley, sprouts (alfalfa and soy), raw Italian squash, Swiss chard, dandelion greens, green cabbage, leeks, nasturtium leaves, Bibb lettuce, green lettuce, skins of unwaxed cucumbers, avocado, horseradish, beet greens, artichokes, asparagus, carrots, tomatoes, mustard greens, corn, sorrel, black radish, pumpkin, and corn.

VEGETABLES AND FRUITS GOOD IRON SOURCES

The sun dried natural fruits are high in iron, apricots being the highest, followed by black figs, prunes, peaches, dates and raisins. Many other foods have a high content of organic iron. Blackstrap and Barbados molasses, raw wheat germ, cooked soy beans, sesame seeds, pumpkin seeds, sunflower seeds, squash seeds, Brewer's yeast, whole barley, dried beans of all kinds, pinto, kidney, lima, lentils, garbanzos, raw and roasted peanuts, almonds, yolk of egg, natural brown rice, dried peas, rice bran, wheat bran, rye, whole grain cereals, and millet.

Many fresh fruits have a high content of organic iron leading the list is blackberries, grapes, cherries, and the juices of these fruits, oranges, peaches, pears, strawberries, blueberries, gooseberries, and raspberries. All these foods will have a higher content of iron if grown in properly fertilized soil (no chemical fertilizer) and absolutely free of poisonous sprays.

Let me impress upon you that your body needs **organic** iron. Not the iron that comes from inorganic sources.

You often hear about a certain well or spring containing large amounts of iron. Yes it does contain inorganic iron. But your body cannot use this inorganic iron — in fact, this iron is dangerous to your body. It can help make all kinds of stones in your vital organs, cement your joints and turn your blood vessels to stone. Again I caution you: **Stay away from inorganic minerals!**

EVERY MINERAL MATTERS

The body contains 19 essential mineral elements, all of which must be derived from food.

Calcium, phosphorus and magnesium are vital for the growth and maintenance of bone; potassium, sodium and chlorine give body fluids their composition and stability.

Calcium, phosphorus and sulphur are essential constituents of all body cells from which all organs and tissues are composed.

Magnesium, iron and phosphorus are parts of enzyme systems concerned with the release of energy from food.

Iodine is important to the thyroid gland which controls growth and the rate at which energy is used.

Copper and iron are needed for the formations of red blood cells.

Other minerals like sulphur and cobalt are used in the synthesis of some vitamins by the body. Zinc is an essential part of the insulin molecule.

Every mineral contributes a unique factor to vitality which is the positive proof of health.

SODIUM, A POWERFUL CHEMICAL SOLVENT

Organic sodium is a powerful chemical solvent and neutralizer of toxic waste products. In contrast, table salt — inorganic sodium chloride — not only is unnecessary but is actually harmful in the body chemistry.

In the animal and human organism, organic sodium has many important functions. In combination with chlorine it is one of the principal constituents of lymph. For the transmission of the electric induction current, which is generated in the nerve spirals by iron of the blood, a salty liquid is necessary (as is shown by the construction of electric batteries). The normal blood serum contains, for this purpose, a comparatively large quantity of organic sodium chloride which favors and sustains the generation and conduction of electric currents.

Moreover, organic sodium plays an important part in the formation of saliva, pancreatic juice and bile. Especially in the bile the dissolving and reducing properties of sodium can be very distinctly recognized in the emulsification and saponification of

55

fats. Organic sodium is a fat fighter. It helps to keep the waxy killer, cholesterol, at normal levels of 150 to 180.

Sodium is essential in purifying the system from carbonaceous waste products. But again let me remind you that sodium is of value to the system only when supplied in organic form, as contained in fruits and vegetables!

BEETS FOR LONGEVITY

Raw beets and celery have the highest amount of organic sodium. I eat them every day in salads, as juices and as cooked vegetables. Several times a week I make a Borscht type of beet soup. Here is the recipe. I hope you like it.

The Bragg Special Beet Soup

1 small onion, minced
3 cups shredded raw beets
1 cup shredded carrots
1 cup diced celery
2 potatoes, diced, skin and all
2 cloves garlic, minced

2 cups shredded green cabbage
3 fresh tomatoes (or 1 cup
    canned unsalted tomatoes)
1 tsp. lemon juice or natural
    cider vinegar
½ tsp. kelp seasoning

    2 tbs. unsaturated oil such as soy, corn, olive, safflower
      ½ cup sour cream, optional

Mince onion and saute in oil until yellow, about 3 minutes. Add 1½ quarts of steam-produced distilled water with vegetables. Simmer about 15 to 20 minutes until vegetables are tender. Season with lemon juice, or pure natural apple cider vinegar. Optional: serve garnished with sour cream.

This is a real natural organic sodium soup and varies but slightly from the original Russian Borscht. As a scientist and researcher in nutrition, I have been keenly interested in the long-lived Russians. I have made a number of expeditions to primitive Russia and have found there people who were amazingly long-lived, some as old as 164 years.

I found that many of them drank only rain water and snow water, thus reducing their chances of having hardening of the arteries. And beets, I learned, were an important food in their daily diet. In fresh water streams they get watercress, which they mixed with raw grated beets. Many of these Russians had never tasted common table salt. Their arteries were flexible and free from inorganic mineral incrustations.

SALT — A SLOW BUT SURE KILLER

Common table salt (inorganic sodium chloride) is both unnecessary and injurious to the human body. It is just like the inorganic minerals to be found in all drinking water except steam-produced distilled water.

Salt can help to form stones in the body. Salt can cause incrustations in the arteries, veins and capillaries. Salt can waterlog body tissues, making them flabby and without skin or muscle tone.

As a general rule, salt users have high elevations of blood pressure. According to medical statistics, the Japanese suffer from the

highest blood pressure in the world, and they are known to be the world's highest salt consumers. My grandfather, who had a massive stroke in front of my eyes at the dinner table, was a heavy user of salt. He put salt on everything he ate, including tomatoes, watermelon, cantalope, celery and radishes — and he ate all the salty foods such as ham, bacon, corned beef, hot dogs, lunch meats, salted popcorn, pretzels, and salted nuts.

Eating such large amounts of salt and salty foods gave him an unquenchable thirst. I have seen my grandfather consume two pitchers of water at one meal. He ate the salt and salty foods and washed these down with hard well water that was full of inorganic minerals. Is it any wonder that his arteries turned to stone?

As previously noted, inorganic salt is indigestible and, when eaten in large quantities, cannot be eliminated from the system. It is therefore deposited in the tissues of the body, and the craving for water develops from the body's attempt to make the salt soluble. Thus the tissues, and the vital organs, become waterlogged. When this condition reaches the heart, we have what is known as congestive heart failure. With hardened arteries and tissues, plus the flabbiness from waterlogging, the heart cannot function.

OVERWEIGHT AND DROPSY

Long before the stage of congestive heart failure is reached, the excessive salt eater suffers many miseries.

The most common of these is overweight and obesity. Statistics show that 65% of Americans are overweight — and not all of this is excess fat. Many times overweight is due to waterlogged tissues. And this overweight problem will continue as long as such people use the salt shaker on their foods, and especially if they partake freely of salty canned fish, salted butter, ham, bacon, lunch meats, salted canned vegetables, salted frozen dinners, salted cheese, salted popcorn and salted nuts.

Like hard water, such a salt-filled diet will also damage the arteries, veins and capillaries.

The kidneys are most severely affected by the salt eating habit. They become weakened and unable to eliminate the large amount of salt, which is then returned to the tissues where, of

----

OVERWEIGHT, OBESITY, DROPSY AND EDEMA. When you are between 10 and 15 pounds above your normal weight, you are considered overweight, above that you are obese. Dropsy and edema means an excessive accumulation of fluid in the tissues, thus causing swelling. When your face, neck, body and the ankles swell, the heart is not functioning correctly.

WHEN YOU ARE HEALTHY — YOU ARE HAPPY!

JOIN THE FUN AT THE "LONGER LIFE, HEALTH
AND HAPPINESS CLUB" WHEN YOU VISIT HAWAII

Paul and Patricia Bragg and some of their prize members of the "Longer Life, Health and Happiness Club" at their exercise compound at Fort DeRussy, right at Waikiki Beach, Honolulu, Hawaii. Membership is free and open to everyone who wishes to attend any morning Monday through Saturday from 8:30 a.m. to 10:30 a.m. for deep breathing, exercising, meditation, group singing and mini health lectures on how to live a long, healthy life! The group averages 75 to 100 per day. When they are away lecturing they have their leaders carry on until their return. Thousands have visited the club from around the world and then they carry the message of health and happiness to their friends and relatives back home. Paul and Patricia extend an invitation to you and your friends to join the club for health and happiness fellowship with them . . . when you visit Hawaii!

course, it must be held in solution by water. This condition produces dropsy, which generally occurs with Bright's disease and cirrhosis.

Dropsy is a common disease in this country. Observe the ankles of the average person, all too often swollen and puffed-up. This condition is sometimes so severe that the ankles must be bandaged before the afflicted person can stand up. In time, this dropsy goes into a chronic state and interferes with the circulation to such a degree that mortification or gangrene sets in, and an amputation becomes necessary.

There are no organs in the body so mercilessly maltreated as the liver and the kidneys. Think of the gallons of water saturated with inorganic minerals which these organs try to neutralize! Not only

drinking water, but water mixed with coffee, tea, alcohol, colas and other soft drinks plus catsup, mustard and other seasonings with high salt concentration. Our poor kidneys and liver! What a terrible beating they take! No wonder most people sicken and die long before their time.

Man does not die: he kills himself by faulty, haphazard living!

MISINFORMATION ABOUT SALT

Using common table salt is one of the widespread, perverted, injurious habits of man. The consumption of salt in the United States now amounts to more than 100 pounds per capita, and it is constantly increasing. The salt companies do a thorough brainwashing job, even telling people they need iodized salt as a prevention of goiter.

Chloride of sodium, or common salt, is an inorganic substance which has caused much confusion in the minds of people for many, many years, particularly in regard to its necessity as an adjunct to our food.

We Constantly Meet Such Erroneous Statements As These:

"It is the only substance which we take into our bodies directly from mineral elements."

"The desire for salt is instinctive with nearly all animals."

"Common salt is one of the most essential of the mineral constituents of the body."

"In hot weather when we sweat a great deal, we lose the salt from our bodies, and we should use large amounts of salt and salt tablets to replace it; otherwise, we will get sick, weak and suffer from extreme exhaustion."

"When salt is entirely withheld from an animal, death from salt starvation ensues."

"Without salt, we would die."

All of these assertions, and many similar ones, are diametrically opposed to the truth. Why should chloride of sodium be an exception to other inorganic minerals?

This homemaker is preparing delicious and nutritious foods for her family without salt.

Salt has been used in the human diet for thousands of years — but not because the human body needs it. Salt was the first food preservative discovered by man, and it is still used extensively in the preservation of nearly all foods, especially meat and cheeses. It is found in canned baby foods, canned soups, canned vegetables, canned fish, prepared cereals, all commercial breads and bakery goods; in fact, it is very hard to find any foods in the super markets that have not been contaminated with salt. Even in this modern age of refrigeration and other mechanical marvels, man has not progressed beyond the primitive stage in the use of salt to lengthen the life of his food — and shorten his own life.

The salt eating habit is not instinctive. It is acquired, as are other health-destroying, life-shortening, unhygienic habits. The taste or craving for salt is artificial, because salt paralyzes the 260 taste buds in the mouth. Like any other addiction, salt creates an unnatural craving by deadening certain of the body's warning signals.

If we choose our food correctly, there is absolutely no true necessity for salt.

The advocates of salt point to the animals who often go miles to so-called "salt licks". I have made a study of the natural "salt licks", and on careful examination found little, if any inorganic sodium.

Cattle, like humans, do need organic sodium. When cattle are fed on herbage grown on soils poor in mineral elements, especially sodium, as on mountain slopes where rains have carried away the most soluble parts of the soil and deposited them in the valleys, they may try to satisfy this lack at an (inorganic) artificial "salt-lick".

An animal's taste buds, just like a human's, can be perverted with salt. Often the salt block is put in the pasture so the cattle will lick it, become excessively thirsty, and consume large amounts of water. As in humans, the result is waterlogged tissues. Consequently the cattleman will gain on waterlogged tissue weight when the cattle are brought to market. Remember, when you eat commercial meat it may be well saturated with salt. Don't make matters worse by adding more salt to it!

Plenty of fresh fruits and vegetables in your diet will supply your body with all the organic sodium it needs — as well as with pure, distilled water. You can find no purer drink than the unadulterated juice of fresh fruits and vegetables and steam distilled water.

INORGANIC WATER TURNS PEOPLE TO STONE

When I was a small boy, my father took me and other children of the family from our Virginia home to Washington, D.C. to see the P. T. Barnum Circus. To a boy off a farm, this was a great event. After seeing the big circus show in the big tent, we visited the "Side Show" tent where all the freaks were exhibited. There were fat men and women, some weighing as much as 600 pounds, the dwarfs, the giants, the bearded lady, the monkey man and others.

But the most fascinating freak to me was the lady who had turned to stone. Here was a woman on a bed, and they could actually drive spikes and nails into her body. She was so full of arthritis and acid crystals that she had no feeling left in her body. She lay helpless and rigid. She could move only her eyes. This lady suffered with complete ankylosis — that is, there was no joint in her entire body that could make a simple movement. All the nerve tissue in her body was paralyzed and dead. The man who explained these freaks said that this lady was born in Hot Springs, Arkansas.

The lady who had turned to stone was a complete mystery to me as a child. But not today! The water in Hot Springs is some of the heaviest water in the United States. I have seen analyses of it, and the concentrations of calcium carbonate, potassium carbonate and magnesium carbonate are very, very high. The poor lady in the side show was a victim of this inorganic water. Her vital organs were not strong enough to help get those inorganic minerals out of her body, so they deposited themselves in her joints.

This was an unusual case, of course. But I have seen many, many cases of arthritics who were complete cripples, absolutely helpless.

There are more than 20 million people living in the United States today who have arthritis in some degree.

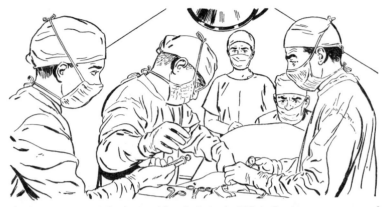

There are nearly eight thousand hospitals in the U.S.A. Surgery goes on around the clock. Many people go into surgery to have painful spurs removed, also bladder stones, kidney stones and gall stones. Will you be next?

BONE SPURS AND MOVEABLE JOINT CALCIFICATION

Every day numbers of people go into surgery to have painful, crippling bone spurs removed and to have calcified deposits removed from the moveable joints. These bone spurs and calcified formations are insoluble deposits that get into the tissues from consuming water loaded with inorganic minerals, salt and uric acid, plus toxic acid crystals from an incorrect diet high in acid. Meat, potatoes, refined flour, white bread, coffee, tea and sugary desserts are all high in acid content. This is the dead-food diet that most people eat — and that, plus hard water, is why there are so many troubles resulting from acid deposits that form bone spurs and crystallized joints.

All diets must be balanced on the acid-alkaline basis. In general, starchy and fatty foods, refined white sugar and proteins are acid forming, while fruits and vegetables (with a few exceptions) are alkaline forming. A balanced diet is one that is about 3/5 alkaline and 2/5 acid.

CALCIFICATION TEST

Make this test to see how much calcification you have.

Stand erect, hands loosely at sides. Now lower the head to the chest and start a slow rolling movement, around and around. Many people can feel the inorganic calcification grating as they roll their heads. This shows that there has been a definite infil-

tration of insoluble minerals and toxic acid crystals in the head
of the atlas, the spine bone upon which the skull rests.

Also test the moveable joints of the body. Do you have stiffness? How limber is your spine? Can you raise your hands over
your head and bend forward with the knees stiff to touch the floor
with your fingertips? Are you limber enough to place the palm of
your hands on the floor as you keep the knees stiff?

Stand with your back to a wall. Move forward about two feet,
then bend backward and "walk" down the wall with your hands.
How far can you go?

How high can you kick?

Do you have cracking in your knees when you do a leg squat?

How flexible are your feet? Do you walk with a spring in your
step?

Do you have a feeling of suppleness and flexibility in your
body? Do you walk and dance with grace and limberness?

Or do you walk on inorganic calcified incrustations that give
you misery?

Don't tell me your stiffness is due to your age. To me that is
just so much "rubbish". You can keep your body flexible by
proper care of your body.

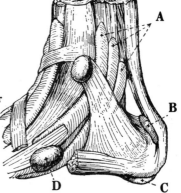

**DEPOSITS OF INORGANIC
MINERALS AND TOXIC ACID
CRYSTALS IN THE HEEL OF THE
FOOT CAUSING GREAT PAIN!**

A. Inorganic minerals deposited under
the tendons.

B. Under the tendon of Achilles.

C. Under the heel.

D. Under the middle foot.

CALCIFIED TOE AND FINGER NAILS

Inorganic minerals, salt and toxic acid crystals can deform
the toe and finger nails. I have seen toes and fingers that were
monstrosities from calcified joints and nails. Big, thick toe nails
like cement that no scissors or nippers could cut — they had to
be filed down with a rugged file. They deform the feet, make
walking extremely painful, and are a hideous sight.

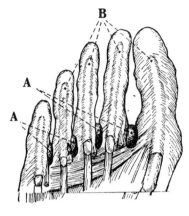

INORGANIC MINERAL DEPOSITS THAT MAY DEPOSIT THEMSELVES BETWEEN THE BONES OF THE TOES (A, B,) CAUSING STIFFNESS IN THE FEET.

BAD POSTURE FROM INORGANIC DEPOSITS

Inorganic minerals and toxic acid crystals are a major cause of poor posture, which brings on all sorts of disturbances by throwing the vital organs out of place, unduly straining some muscles while weakening others, impairing circulation, breathing and elimination.

Just stand on any busy city corner and watch the people go by. What a miserable sight most of them are! People whose feet

WALKING POSTURE

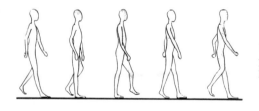

Walking posture. Always prepare a new base before leaving the old.

LIFTING POSTURE

Lifting weight. The weight of the baby is held close to the center of gravity directly above the pushing force.

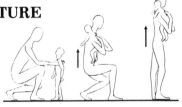

GOOD AND BAD WAYS TO:

Sit

Walk

Lounge

are so loaded with inorganic calcification that they simply lift their feet up and put them down; the spring absolutely gone from their step. Many walk like ducks with the toes pointing outward to the sides. Others are stooped and bent out of shape. Some walk with no knee action. You see those whose steps are unsteady because their joints are so cemented; while others are so out of balance that they sway from side to side as they hobble along. Heads are carried too far forward, throwing the body off its balance.

And watch them as they try to sit down. They simply slump into the chair, giving the lower back a shock.

THE CURSE OF AN ACHING BACK

By the time they reach 40, most people are plagued with low back pains. When they bend over it is positive agony. The whole lower spine is cemented with inorganic calcification.

By the time a person is 40, he has worn out much of the cartilage that acts as a cushion to the spinal bones. This is a painful condition, which gets progressively worse on a continuing diet of chemicalized, inorganically mineralized water and a preponderantly acid crystal-forming diet.

THE PARADE OF THE LIVING DEAD

Just remember that chemically, physically and mentally we are what we eat and drink. Because most humans are totally ignorant about what to put into their bodies, very few escape being one of the living dead. So many people, after their twenties, do not know what healthy, vibrant, youthful living is. They drag themselves through life, relying on some kind of medication to keep them going. They need a "pep" pill to keep them going during the day and a sleeping pill to put them to sleep at night.

67

CHECK YOUR MATTRESS

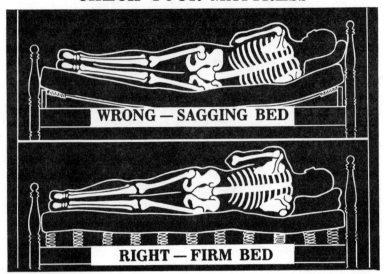

WRONG — SAGGING BED

RIGHT — FIRM BED

During sleep, you recharge the battery you ran down slowly during the day. The right kind of mattress is important. It's better to sleep ON the mattress than IN it.

Mankind is sick and growing sicker year by year. Throughout the whole of recorded history, man has suffered with a variety of miseries — a great many of which can be directly traced to hard, inorganic mineral water.

In a museum in Milwaukee I saw the backbones (spines) of American Indians, who lived in Wisconsin over a thousand years ago, with calcification that showed they were victims of arthritis. These Indians drank the Lake Michigan water, which is heavily saturated with inorganic calcium carbonate and other inorganic minerals. Their spring and river waters were no

Early civilizations suffered from hard water.

better. One thing they did not have to worry about, however, was the harmful chlorinization and fluoridation of their drinking water which would have added to their already burdened bodies.

Even the mummies of ancient Egypt, some 2500 years old or more, show the ravages of arthritis and other diseases due to drinking Nile River water, which is heavily saturated with inorganic minerals.

So, as you can see, even the most primitive of people, living under the most natural conditions, suffered and died long before their time. Inorganically mineralized water is surely the universal drink of disease and death.

Every time a person turns on the water faucet and drinks water that has been chemicalized with chlorine and is saturated with calcium carbonate and other inorganic minerals, he is jeopardizing his health and his life!

THE "FAD" OF DRINKING SEA WATER

From time to time during the past 60 years, some so-called "health experts" have advised drinking sea water to get the minerals which the body requires. They give you the argument that billions of tons of top soil is washed into the ocean every year, and that these minerals from this rich soil can be drunk by the human body to gain more health.

Nothing could be farther from the truth. Yes, the ocean is a vast storehouse of inorganic minerals. But again I must positively state that the human body cannot utilize any inorganic mineral — whether it comes from a well, spring, river, lake or ocean. Then, too, ocean water has a very high concentration of sodium chloride (common salt) which cannot be used by the body chemistry.

Don't drink sea water, no matter what you have read! Sailors and shipwrecked people have tried it many times, and it sent them raving crazy and to an agonizing death.

The best service a book can render is, to impart truth, but to make you think it out for yourself.

— **Elbert Hubbard**

SEA KELP — RICH MINERALS FROM THE SEA

Now, when you eat sea plants such as kelp and seaweed, you are following the rules of scientific nutrition. The sea vegetation converts the inorganic minerals of the sea into organic minerals, and it is healthily profitable to eat any kind of sea vegetation. On my salads and other food I sprinkle sea kelp. It gives the food a tangy flavor and at the same time furnishes the body with a large amount of iodine. In addition to this kelp seasoning of my food, I take a 5-grain tablet of kelp every day to be assured that I get my daily ration of that all important organic mineral, iodine.

DISFIGURING BROKEN CAPILLARIES

Among the many manifestations of inorganic calcification are broken facial capillaries.

Study people's faces. Look closely on the cheeks, around the nose and on the chin, where you will often see the smallest blood vessels, slender as hairs, showing near the surface of the skin. When these tiny capillaries become incrusted with inorganic minerals, they expand in size and often rupture, making purplish or reddish blotches. Blocked by inorganic minerals and no longer able to handle the circulation of the blood, except perhaps to a small degree, these broken capillaries not only give the face a grotesque appearance but are often quite painful.

COLD FEET AND COLD HANDS

Many people of all ages suffer with poor circulation, in most cases due to or aggravated by inorganic mineral incrustations in the arteries, veins and capillaries that constitute the blood circulatory system.

I have frequently shaken hands with people whose hands were ice cold even on warm days. Many people also suffer with extremely cold feet, especially in cool weather. By age 60, most people have patches of small, blue, broken and expanded veins around their feet and ankles, giving an appearance of blackness, often looking dirty, even just after a bath.

Poor circulation is first and most critically noticeable in the hands and feet, because the blood has farther to go from the heart to reach these extremities. When the pipes of the body become clogged and obstructed, the blood has difficulty in getting through. Instead of coursing through the capillaries of hands and feet in a warm, healthy stream, it trickles through, barely able to bring nourishment, without imparting warmth.

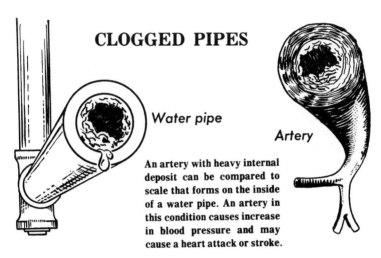

CLOGGED PIPES

Water pipe

Artery

An artery with heavy internal deposit can be compared to scale that forms on the inside of a water pipe. An artery in this condition causes increase in blood pressure and may cause a heart attack or stroke.

The entire body, of course, is affected when the pipes of the circulatory system are clogged. People with poor circulation find it difficult to keep warm in cold or even cool weather. Their homes are over-heated, and they must bundle up in sweaters, coats and other bunglesome clothing when they go outside.

Every sick person has, more or less, a circulatory system that is operating on a very low level — chiefly due to plugged-up pipes. And to repeat, the main source of these incrustations is drinking water saturated with inorganic minerals. Those who drink distilled water and juices of fruit and vegetables are helping to keep their circulatory system clean and healthy.

EXERCISES FOR HEALTHY FEET

A few simple exercises each day such as the following will help to keep them in good condition and improve circulation.

1. Raise the weight of the body on the toes.
2. Grasp with the toes, preferably picking up a pencil.
3. With the knees crossed, extend the foot completely and then fully flex it.

71

4. In the same position, rotate the foot clockwise as far as possible several times, and then counter clockwise.

5. Sit on floor with soles of feet together and hands grasping ankles; pull heels and toes alternately apart.

6. Stand with feet parallel, three to four inches apart, slightly bend the knees and turn them outward; keep the feet flat on the floor.

Tired and aching feet may be helped by lying on the back and putting a pillow under the feet to reduce the congestion or by holding the feet in the air even for only a short time. Rest of the feet is as important as rest for the body.

7. Walk barefooted on soft grass or sand anytime the opportunity arises. Your feet love the earth contact, plus the exercise and the step up in circulation you acquire.

8. Self-foot massage is a wonderful treat to indulge in while watching TV or listening to the radio — rotate and work, squeezing each toe and then apply pressure massage to the bottoms of the feet.

Tired and aching feet may be helped by lying on the back and putting a pillow under the feet to reduce the congestion or by holding the feet in the air even for only a short time. Rest of the feet is as important as rest for the body. (My book **"Building Strong Feet"** is a complete guide for strengthening and building strong feet. See back cover for details.)

HEAD NOISES, RINGING IN EARS

Many humans are plagued by head noises, ringing and pounding in the ears. Every waking hour is a torture as these head noises wear out one's nerve energy. Even in sleep, they have bad dreams about these noises. These head noises torment them 24 hours a day.

The blood vessels — arteries, veins and capillaries — in the delicate pipes of the ears have become hardened and obstructed with inorganic mineral incrustations from hard water, as well as deposits of toxic acid crystals from unbalanced diets, plus uric acid. This condition produces the head noises, ringing and pounding in the ears and buzzing sounds.

In time, the blood vessels of the ears become so clogged that the person gradually goes deaf. Thousands of people go deaf

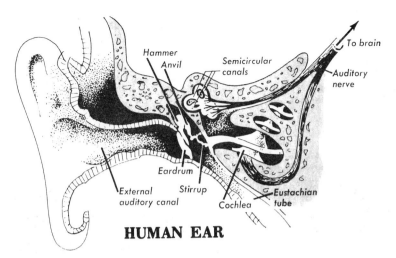

Hammer

Anvil

Semicircular canals

To brain

Auditory nerve

Eardrum

External auditory canal

Stirrup

Cochlea

Eustachian tube

HUMAN EAR

A diagramatic representation of the parts of the human ear.

every year. For a time they can get some relief from a hearing aid, but many lose their hearing completely.

EFFECT ON THE EYES

Inorganic minerals, toxic poisons and uric acid also have a degenerating effect upon the eyes.

Your eyes are among your most important physical possessions. They are often described as the mirror of the soul, the mind and the thoughts. It is true that your eyes often reveal your innermost feelings. Their lustre changes under psychological influences, such as fear, love, hatred, as well as under all physical malfunctions. Without your eyes, you would live in total darkness. Many do.

Now let us examine for a moment this wonderful mechanism, the eye. The eyeball is an almost spherical body with a mirror at the back portion known as the retina. The body of the eye is made up of a transparent jelly-like substance. Through the eye runs the optic sensory nerve. This is a cranial (head) nerve, a direct part of the central nervous system.

In the front of the eye there is a crystalline biconvex lens, which is more convex behind the cornea.

The white portion of the eye which you see is the fibrous coat surrounding the central, colored part — varying in shades of

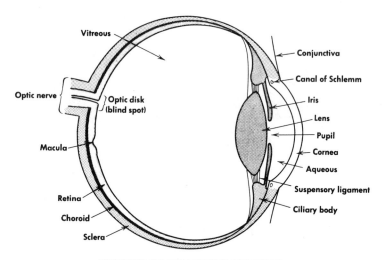

SECTION OF THE RIGHT EYEBALL

brown, blue, hazel or gray — known as the iris. The iris serves as a diaphragm, controlling the amount of light which enters through the pupil, the black spot in the center of the eye. When the outside light is bright, the iris contracts the pupil to a small dot. When the outside light grows less bright, the pupil is expanded in proportion.

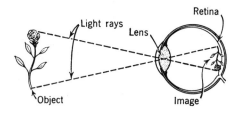

The human eye is very much like a camera. Light rays enter the eye, cross in the lens, and focus on the retina.

NATURAL FOODS HELP YOUR EYESIGHT

So, as you see, the eye is a most delicate mechanism. All the blood vessels which bring nourishment and oxygen to the eye are very small capillaries.

Now year after year, people drink chemicalized and inorganically mineralized water and absorb the toxic and uric acid from their daily food. Just as in the limestone caverns the stalactites and stalagmites are formed drop by drop, so — drop by drop — the blood brings deposits of inorganic minerals and toxic acids into the blood vessels of the eyes. In these delicate capillaries incrustations are formed. Glasses are prescribed. After a time stronger glasses are necessary. Then the vision starts to fail, and in many instances a person is left in total darkness.

Many people start to panic when they start losing their wonderful vision. They try treatments of all kinds and descriptions; some have operations, but the sight gradually fades away.

Just remember you have three vicious enemies to the eyes: inorganic mineral water, toxic poisons from acid-forming food, and uric acid from a diet too heavy in animal proteins.

BUILD CLEAN BLOOD

The blood holds the key to our health, our vitality, our youthfulness — and our life! **Keep the blood free from inorganic minerals and toxic acids!**

Every 90 days we build a brand new bloodstream. We can live and regain health by the reversal program if we will be careful of the kind of water we drink and the kind of food we eat. So, today you can discard the materials which build an unhealthy bloodstream and start building one that is going to give you a Painless, Tireless and Ageless Body! It's all in your hands! With the knowledge given in this book you can start getting more

WHITE BLOOD CELLS

A   B   C   D   E   F

White blood cells: b, lymphocyte; c, basophil; d, neutrophil; e, eosinophil; f, monocyte. A red blood cell (a) is included for a comparison of sizes.

youthful as you live longer. You must be the absolute master of what goes into your body in the way of food and drink. Flesh is dumb. You can put any kind of food and drink into your body and, in most instances, your body will accept it. That is the reason you must read this book several times and plan at once a Program of Clean Blood Building.

Of course, we all know what blood looks like, a somewhat thickish red fluid that we see whenever the skin is even slightly broken. These tiny oozings of blood come from very minute blood vessels which supply the skin everywhere in the body.

**If this child were to live on a 100% Health Program
it would have a Long Life and enjoy a Painless,
Tireless and Ageless Body.**

Blood, your river of Life, is the fluid which carries oxygen and
nutrition to all cells of the body and tries to take away poisonous
substances. The trouble is that the average person is pouring inorganic minerals and toxic poisons into the body so fast and
furiously that the blood finds it impossible to purify itself.

Nothing could be more important than this "life blood" of
ours. If we do not get enough oxygen and nutrition, we will die;
and if poisonous materials are not removed, we will die.

And that, I am sorry to say, is why we die long before our time.
We do not take the time to learn how to get more oxygen into
the body; we do not take the time, or we are not interested in
good nutrition. Death comes from accumulated poisons that actually poison and clog up the blood, blood vessels and nerves;
inorganic minerals which become a burden to the body; the
heavy concentration of inorganic salt, fat and vicious toxic poisons — these are the killers! Years are not your enemies. It's
what you put into your body that does the terrific damage to
health and life.

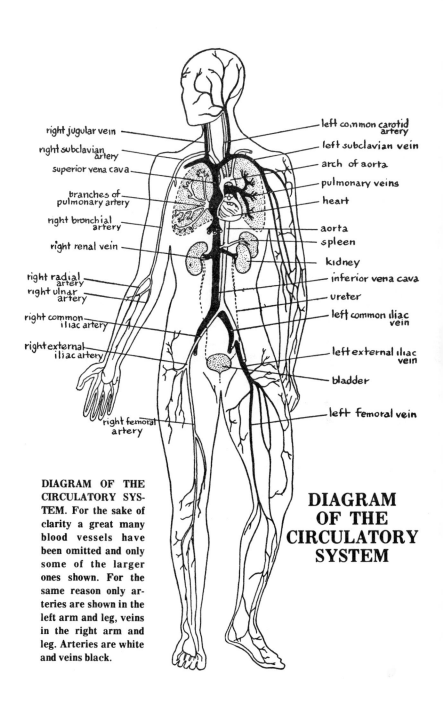

right jugular vein

right subclavian artery

superior vena cava

branches of pulmonary artery

right bronchial artery

right renal vein

right radial artery

right ulnar artery

right common iliac artery

right external iliac artery

right femoral artery

left common carotid artery

left subclavian vein

arch of aorta

pulmonary veins

heart

aorta

spleen

kidney

inferior vena cava

ureter

left common iliac vein

left external iliac vein

bladder

left femoral vein

DIAGRAM OF THE CIRCULATORY SYSTEM. For the sake of clarity a great many blood vessels have been omitted and only some of the larger ones shown. For the same reason only arteries are shown in the left arm and leg, veins in the right arm and leg. Arteries are white and veins black.

DIAGRAM OF THE CIRCULATORY SYSTEM

THE WORLD'S GREATEST HEALTH SECRET

The "secret" of health lies in Internal Cleanliness! To be 100% healthy, a body must be absolutely free of deposits of inorganic minerals that come from drinking city tap water, water from lakes, rivers, wells and springs. Inorganic minerals contaminate the human body, form incrustations that clog and obstruct the human pipe system and impair the vital organs.

The body needs absolutely pure H_2O. This water is at its best when it comes from raw, organically grown fruits and vegetables. This is life-giving water — water that has been charged with solar energy, health building vitamins, organic minerals and the marvelous enzymes. Enzymes may help you build up resistance to any ailment, may help flush out the accumulated deposits of inorganic minerals, may help you dissolve the toxic poisons that have buried themselves deep in your tissues and vital organs.

FRESH JUICES — THE MAGIC CLEANSERS

The raw juices of fruit and vegetables are internal cleansers and blood purifiers. This is what we call "the waters of perpetual health and youthful-ness". The rays of the sun send billions of at-oms into plant life. We can use this solar energy to attain vigorous health, unlimited vitality and physical endurance.

This sun energy in fruits and vegetables can fight the accumulation of inorganic minerals and toxic poisons you have allowed to be deposited in your body. Fruit and vegetable juices are the natural detergents for the human body.

Try to get a quart or more of solar energized (fresh squeezed) fruit and vegetable juices into your body every day.

Go to your Health Store and purchase a juicer and blender. Both of these appliances are important in your program of ridding the body of inorganic minerals and toxic wastes. It will be probably the best investment you ever made in your entire life. With the juicer you can have many varieties of juices. Carrot, celery and raw spinach is a wonderful combination.

Carrot, beet and celery juices make a rich organic sodium cocktail. Apple and cucumber juice is a great health cocktail. Green pepper and tomato juice is a real internal cleanser. Raw spinach and watercress will flood your bloodstream with organic iron. Parsley and carrot juice is a delicious and healthful combination. Cabbage juice, (Stanford University Medical School discovered this helps heal ulcers.), onion juice, garlic juice, pea-pod juice, turnip-top juice, radish juice, lettuce juice, kale juice, dandelion juice and endive juice are all packed with solar energy, vitamins, organic minerals and enzymes.

Fruit juices play an important role in building a clean body and bloodstream. Apple juice, pineapple juice, cherry juice, blackberry juice, orange juice, grapefruit juice, prune juice, apricot juice, strawberry juice; these are the true nectar of the gods.

LET NATURAL FOOD BE YOUR MEDICINE

What does food really do in the human body? What relationship does it have to long life, vigorous health — and to disease, misery and physical suffering? And how can it be of influence in cleansing the body of inorganic minerals and toxic poisons?

These are questions about which one must have a fundamental and profound understanding before one can fully appreciate the role diet plays in the maintenance of living processes, the prevention of human misery, the restoration of health and the prolongation of life.

A balanced diet gives the body nourishment, energy and power. A balanced diet is made up of 3/5 raw fruit and vegetables; 1/5 protein, either animal or vegetable; 1/3 of 1/5 natural sugars such as honey; 1/3 of 1/5 natural starches such as whole grains; brown rice; and 1/3 of 1/5 unsaturated fats such as corn oil, olive oil, soya oil or safflower oil.

This balanced diet puts your body on the alkaline side, and helps keep the body internally clean at all times.

Let food be your medicine, and medicine be your food.
— Hippocrates

EAT ONLY
NATURAL
LIVE
FOODS

AVOID THESE "FOOD-LESS" FOODS ENTIRELY

To keep your body internally Clean, Avoid These Dead, Devitalized so-called foods:

1. Refined white flour and its products, mushy cereals, dry supermarket cereals, and other refined grain foods.
2. Fried or mashed potatoes.
3. Refined white sugar and its products.
4. Coffee, tea, alcohol, cola and other soft drinks.
5. White rice, pearled barley.
6. Salt. Salt has no place in a balanced diet. Read labels, and if "salt" is listed as an ingredient, keep away from it. This includes:
    a) Processed cheese, salty canned vegetables, salted canned fish.

b) Processed meats such as hot dogs, luncheon meats, ham, bacon, corned beef — all loaded with concentrated salt solutions.

When not hungry, eat nothing!
Make your body earn its nourishment by vigorous activity!

RE-EDUCATING YOUR 260 TASTE BUDS

It will take some willpower to go through the transition from eating dead foods to eating live foods. For a while there will be a craving for the unhealthy foods which you have probably eaten all your life. But if you will be positive in your selection of natural foods, the old desire for devitalized foods will leave you.

In time, you would not think of insulting your body with the refined, processed and manufactured foods of civilization. You will be **reborn!** You don't have to be half-alive and sick!

And you will find an added pleasure in enjoying the true tastes of the live foods you eat, which you will be able to discern when your 260 taste buds have recovered from salt paralysis and come alive again.

EASY ON MEAT

Meat is a rich source of protein. In fact, nutritionists call it the Number One Protein. Those proteins coming from the vegetable kingdom are called Number Two Proteins.

Meat, however, is also a major source of uric acid and cholesterol, both harmful to your health. Having spent some 65 years in the research of scientific and natural nutrition, I have reached the following conclusions in regard to eating meat:

If you are going to include meat in your diet, it should not be eaten more than three times a week.

In my opinion, fresh fish is the best of the flesh proteins. Next comes chicken and turkey — but never eat the skin, which is heavy in cholesterol. Third place goes to lamb and beef.

I do not believe people should eat pork or pork products of any kind. The pig is the only animal besides man that develops arteriosclerosis or hardening of the arteries. In fact, this animal

is so loaded with cholesterol that in cold weather, unprotected
pigs and hogs will become solid and
stiff, as though frozen. Also, this
animal is often infected with a dangerous parasite called trichinosis.

As stated before, ham, bacon,
luncheon meats and hot dogs are all
preserved with high concentrations of
inorganic sodium (salt) solutions and

**Garbage-fed hog. Ingestion
of meat from such an animal
may cause trichinosis.**

other harmful preservatives. In my opinion, they are dead and
embalmed and can play no part in good nutrition!

 DRINK ONLY DISTILLED WATER

Outside of the fruit and vegetable juices, I drink no other water
except **steam produced distilled water.** Today, in this polluted
and poisoned world, distilled water is the purest water on the
face of the earth. It contains no solid matter of any kind. It is
made solely of the two gases, hydrogen and oxygen. There are no
minerals in it, organic or inorganic. It can be used as drinking
water to cook with, and water to put in electric steam irons and
batteries.

When distilled water enters the body, it leaves no residue of
any kind. It is free of salt and sodium. It is the most perfect water for the healthy functioning of those great "sieves", the kidneys. It is the perfect liquid for the blood. It is the ideal liquid
for efficient functioning of the lungs, stomach, liver and other
vital organs.

Why? Because it is free of all inorganic minerals. It is so pure
that all liquid drug prescriptions are formulated with distilled
water.

Let no person tell you that distilled water is dead water! Of
course, fish will not live in distilled water. Fish require vegetable growth in water, and vegetable growth needs inorganic
minerals to live.

---

*The unexamined life is not worth living. It is a time to re-evaluate
your past as a guide to your future.* —Socrates

---

*Dear Friend, I wish above all things that thou may prosper and
be in health even as the soul prospers.* —3 John:2

WHAT ABOUT RAIN WATER?

Rain water is ideal distilled water — but today our air is so polluted that it poisons and contaminates this natural water.

It is a theory of mine that the people mentioned in the Bible who lived to reach fantastic ages drank only rain water. Rain water is distilled water from the clouds.

Today, however, we live in the age of pollution when even the rains from heaven are polluted. Such goodies as Strontium 90 from our atomic bombs turn rain water into a deadly poison. From our industrial plants vicious poisons are sent into the air — sulfur dioxides, lead, carbon monoxide and hundreds of other pollutants.

So, in our present civilization, rain water is out of the question. To live in this poisoned world; to survive, and to save ourselves from another kind of destruction — i.e., the complete solidification of the brain structures — we must drink only distilled water.

We do not want our brain arteries and other blood vessels to turn into stone. You see this condition every day in prematurely old people in deep senility. Many times you hear the word "fossil" to describe the prehistoric remnants of animals who lived on the earth ages ago. Yet when you drink a glass of ordinary water, the process of fossilization has already begun. When a person dies of hardening of the arteries, he has reached the ultimate end.

So many times I have heard someone say something such as, "That old fossil John Smith died last night with hardening of the arteries."

Although the remark was crude, it was truthful.

If in this life we escape the deadly degenerative and infectious diseases, we are always haunted by that great killer of mankind, "hardening of the arteries".

Every man is the builder of a temple called his body . . . We are all sculptors and painters, and our material is our own flesh and blood and bones. Any nobleness begins at once to refine a man's features, any meanness or sensuality to imbrute them.

— Henry David Thoreau

STAGES OF ARTERY HARDENING

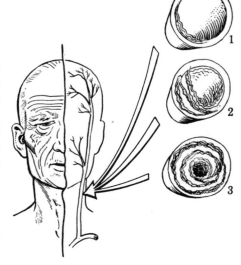

This drawing shows three stages in the hardening of blood vessels in the brain. As the flow of blood becomes slower, clots may form and completely close a vessel.

HOW TO FIGHT HARDENING OF THE ARTERIES

Be determined that you are going to drink only pure distilled water. If you cannot get it from a water company, try the drug store. As a rule drug stores carry distilled water for people who have heart trouble or strokes — but don't wait until that happens to you.

If you cannot find distilled water for sale anywhere, then purchase a small still and distill your own water. You may say that is a lot of trouble — but it's not nearly so much trouble as when your arteries start to harden and your body is slowly starved for want of oxygen. Remember, the blood carries oxygen to all parts of the body. And if the arteries become incrusted with inorganic minerals, you are in for great suffering. Oxygen starvation is a terrible physical problem.

There are no more important ingredients of a properly constituted diet than fruits and vegetables, for they contain vitamins of every class, recognized and unrecognized.
—Sir Robert McCarrison

God sends the food, man by refining and processing foods destroys its nutritional value. Eat only God's natural foods.—**Patricia Bragg**

Patricia Bragg drinks distilled water on world tours.

DURING OUR WORLD HEALTH CRUSADES

During our many Bragg Health Crusades which carried us around the world . . . most often we were able to find distilled water . . . but when not — and if in doubt about the water supply — we would go without the water and for short periods let the fresh raw vegetables and fruits supply us with their pure water. I always have a hand orange juicer with me, for this is the only way I enjoy fresh orange juice.

THE GREAT WATERMELON FLUSH

There is nothing like a watermelon flush to dissolve and eliminate inorganic minerals from your body.

As a youth, I had a history of drinking exceptionally hard water, loaded with inorganic minerals. The crystals of that hard water incrusted the pipes of my body, and when I learned the truth of the great damage it could do to me, I started to experiment with fruits and vegetables to find out which one had the greatest dissolving power. It was a long search, but at last I found it. Watermelon and watermelon juice did the trick.

Several times a year I go on a watermelon flush. That is, I eat nothing for a week or ten days but watermelon and watermelon juice. Every morning I take a sample of the first urine I void the very first thing in the morning. I seal it tightly, date it and put it on a shelf for six months to a year. As it breaks down, the inorganic minerals, being heavier, settle on the bottom of the bottle. Being a biochemist, I thoroughly analyze these substances and find calcium carbonate, magnesium carbonate and other inorganic minerals.

"Our Prayers should be for a sound mind in a healthy body."
— Juvenal

Life is like a gun. It can be aimed in only one direction at a time.
Make your aim — health!
— Paul C. Bragg

That is why I take a watermelon flush several times a year. I also eat watermelon all during its season. Many times I have paid high prices during the winter months for watermelon shipped from warm climates, but I consider that just good health insurance. On an average day in the summertime, I drink as much as a quart of watermelon juice.

The result is that I have the blood pressure of a man of 25: it is 120 over 80, and my pulse is around 60. I have my heart and arteries examined by a heart specialist every year. So far, there is not the slightest sign of any hardening of the arteries. I can jog, run, swim, ride my bike for miles. I have supple, flexible arteries and blood vessels. There are no big, bulging veins on my forehead, such as I have seen on men a third of my calendar years. I can stand on my head for 30 minutes at a time with absolutely no dizziness or other reaction. I enjoy perfect hearing and vision.

I am constantly fighting the worst enemy, "hardening of the arteries", and I expect to remain the victor for many, many more years.

MY PREDICTION

Mine is only a small voice in the wilderness on this matter of drinking only distilled water to save humanity from the dangers of water filled with chemicals and inorganic minerals.

I have lived a long life. During this time I have seen my relatives, my personal friends, as well as some very fine animals, die of fossilization.

I believe I am a hundred years ahead of my time in my theories on the dangers of inorganic water.

But someday humanity will recognize the dangers in ordinary water, and all water used in homes will be completely steam distilled. It will be the greatest health advancement this world has ever known.

No matter how much a person controls his eating habits, no matter how much juice of fruits and vegetables he drinks, no matter if he lives on a raw food diet, vegetarian diet or so-called modern scientific diet — as long as he continues to drink spring water, well water, river or lake water, he is going to fossilize himself. In the past 60 years I have met all the greats in the field of nutrition, natural healing, etc., but they all drink the deadly inorganically mineralized water — and only a very few of them have attained an extra long life.

The same with the great athletes of the past 60 years. They had their day in the sun, drank the deadly inorganic mineral water and died about the same age as a non-athletic person.

I was personally acquainted with Bill Tilden, the greatest tennis player of all times. In his prime no man in the world could defeat him. But he would not listen to my weak, small voice when I told him of the viciousness and deadliness of inorganic drinking water. He said that everybody else was drinking ordinary water, so he saw no reason why he should stop. All his athletic strength and prowess did not save him from having a massive coronary, and he died before he was sixty. The autopsy showed the arteries of his heart were like stone.

Sandow, the greatest strong man of all times, was a friend of mine. When I would visit him in his studio in London, England, he would flex his muscles and tell me how powerful he was. But he drank London tap water, and at 58 he also had a massive coronary. His great strength, his bulging muscles did not save him.

Man is as old as his arteries. There is absolutely no way to get around this statement.

In the course of the Korean War, approximately 300 young

American soldiers were killed during a short period. Autopsies were performed on all of them. And what do you think was revealed? That these young men all showed signs of hardening of the arteries. This scientific study is on record. Here were young men under 23, in the so-called prime of life, showing degeneration of the arteries of the body.

LIFE EXPECTANCY — LIFE SPAN

We are told that a male child born today has a life expectancy of 72 years, and a female about 78. What is the actual life span of the average American? A male child about 65 years, and a female child about 69 years.

So we can plainly see that when a male reaches 33 years he has lived about half his lifetime, and when a female reaches 34 years she has lived about half her lifetime. Sad facts ... some of them never reach their expected lifespan.

The United States and other western countries are experiencing an epidemic of heart disease and also a rise in cancer cases.

TIME

I have just a little minute,
Only sixty seconds in it,
Just a tiny little minute,
Give account if I abuse it;
Forced upon me; can't refuse it.
Didn't seek it, didn't choose it,
But it's up to me to use it.
I must suffer if I lose it;
But eternity is in it.—Unknown.

Face this startling fact: The chances are better than 2 to 1 that, directly or indirectly, the average American adult will die of some form of heart disease.

You as an individual can reverse this shocking fact today by discarding ordinary inorganic mineral water, stopping the use of salt, reducing the amount of saturated fatty foods — and start drinking distilled water and eating a balanced natural diet.

EXERCISE IS NECESSARY FOR HEALTH

For youthful arteries, exercise is essential. To live long it is necessary to build up your cardio-vascular endurance, along with your program of keeping the arteries soft and flexible.

The first step is to get more oxygen into the body to help dissolve the incrustations on the arteries. Any physical activity that helps you bring oxygen into the body is going to help extend your life. Get out and jog, swim, ride a bike, or take a brisk two-mile walk.

KEEP YOUNG BIOLOGICALLY
WITH EXERCISE AND GOOD NUTRITION

Why grow old? Why not grow biologically young as you live longer? Don't let parts of your wonderful body become fossilized by drinking inorganic mineral water. Aim at perfection of your body.

BE A RULER OF YOURSELF

Don't drink chemicalized and inorganically mineralized water. Don't eat foods your judgment tells you should not be eaten. Be stronger than your physical cravings. Do not give way to any temptation, and you will soon find that the inclination will vanish. The really healthy man is always able to rule himself. The greatest triumph is the victory over self.

Don't be afraid of being called a "crank" or "faddist" because you wish to live abundantly, through regulating your drinking and eating habits. Probably there has never been a change that brought blessings to humanity, which was not ridiculed at first.

You can always remember that you have the following good reason for sticking to your health program:

- The ironclad laws of Nature.
- Your common sense which tells you that you are doing right.
- Your aim to make your health better and your life longer.
- Your resolve to prevent illness so that you may enjoy life.

- By making an art of life, you will be young at any age.
- You will retain your faculties and be hale, hearty, active and useful far beyond the ordinary length of days, and you will also possess superior mental and physical powers.

LIFE'S GREATEST TREASURE IS RADIANT HEALTH

"There is no substitute for Health. Those who possess it are richer than kings."

COMFORT, SECURITY AND HAPPINESS FOR YOU

Can you think of any greater comfort than the confidence that you will never be the victim of inorganic minerals, toxic poisons and uric acid? That you will be able to keep these killers out of your body?

Would it not be a wonderful comfort to you to have this great anxiety removed from your life? How glorious to feel the positive conviction that you can live about as long as you like, and that you can teach your children and other relatives and friends how to live in comfort and security against all human miseries and untimely death!

Surely the removal of this stupendous anxiety from our minds seems "too good to be true". Yet this delightful state of affairs is not only possible but is easily attainable by anyone willing to apply the principles enunciated in this little book.

I say to you in all earnestness and with the strongest conviction that I firmly believe all manner of illness and misery is wholly of our own making; and that it is easily preventable by keeping away from all chemicalized and inorganic mineral water; by keeping away from salt in any form; by practicing healthy food habits and by having a definite program of exercise.

PREVENTION BETTER THAN CURE

We live continually under the depressing fear of sickness and death, although we seldom realize it.

We would even indignantly deny this imputation and insist that we are not afraid to die — although most of us are willing to admit the fear of developing some deadly physical condition that may make us a less efficient machine than we are now.

We ever fear those things that we do not understand, for the very reason that we have learned to expect catastrophe from unanticipated sources.

If we fully understand what disease is and how it originates; if we are familiar with the only avenues through which this can come upon us — then what have we to fear except ourselves?

To prevent disease is to cease the daily cause of disease. And the cause of disease, as previously discussed, is the clogging of our bodies with dead chemicals and inorganic minerals, salt, and the increasing amounts of toxic, acid end-products of digestion and metabolism — conditions which we can control.

So, if disease will recover by a reversal of our wrong habits of living, will not these same good habits prevent disease in the first place? Prevention costs nothing, but it does save a lot. If one is of sound mind, it must seem that the only sane method is to avoid habitually the well outlined causes of disease, without waiting for it to develop.

Prevention is the only way you may be helped to feel youthful, vital, energetic and virile.

If we are to prevent disease, we must have a rudimentary knowledge of our body. Our body consists of millions of cells bathed in an electrolytic solution consisting of calcium, magnesium potassium, sodium, phosphorous, chlorides and sulphates, with trace elements of copper and zinc. These are all organic minerals. The body cannot use inorganic minerals for building its cells.

These electrolytes are held in solution by water, which makes up 80% of the body. For this reason, we can live without food for long periods, but we can live only about 72 hours without water.

So you can see how important it is that the body not only gets sufficient water, but that the right kind of water is absolutely necessary.

Of course, the vast majority of people drink ordinary water with its chemicals and inorganic minerals. The body can make no selections for itself, but must accept what you put in it. When you give the body water that is heavily loaded with chemicals and inorganic minerals, the body has to do something with these poisons. It therefore stores them in your arteries, veins, joints, eyes, ears, nose, throat, gall bladder and vital organs.

WHAT PUNISHMENT OUR BODIES CAN ENDURE AND STILL SURVIVE!

For many years the body seems to handle the situation, because the body is such a wonderful instrument that it can take a great deal of poison and still function. But the day finally arrives when the stones you have loaded into your body begin to give you trouble — real trouble: pain, suffering, misery, agony.

Those people who have laughed in Nature's face now cry out in pain, "Help me!" "Save me from my terrible suffering!" These are the people who want a Cure. A cure? No one can cure you of anything. Only the basic biological functions of the body can perform a cure.

Don't wait until pain strikes to start taking care of your wonderful body. It may be too late then. Today is the day to outline a health program for yourself and live by it faithfully.

You get only one body during your lifetime, and if you are to live in health and freedom from suffering, you must faithfully follow the Health Laws of Nature. These are good, kind laws. Nature wants you to have a painless, tireless, ageless body. It is your birthright to feel the thrill of joyous living every day of your life.

By following Nature's great, good laws you can awaken one beautiful morning and discover the feeling of enjoying health and happiness. Gone is your chronic fatigue, gone your headaches, gone all your aches and pains. You feel new vitality surging throughout your entire body. You have a spring in your step, a sparkle in your eyes, and the glow of health in your skin. You have found the greatest and most precious treasure in the whole world — radiant, glorious health. You now sleep deeply like a baby. Physical and mental health are now yours.

The Girl with the Golden Glow

It starts inside of her, a gladness about being alive and healthy, and it can't help shining out of her face.

PERPETUAL YOUTHFULNESS CAN BE YOURS

Longfellow says: "In youth the heart exults and sings" — thus conveying the idea that beyond certain prescribed years the heart does not exult and sing.

Youth, in its broad sense, should not refer to years but to a state of being. It is really a matter of one's own choice, if one is born normal, as to when youth shall end and "middle age" begin. Some people can be truly referred to as youthful although they may have seen fourscore years and ten.

Every normal person really desires to remain youthful, but all do not want to pay the price. The price of prolonged youthfulness is the performance of certain acts.

These acts are: staying away from chemicalized drinking water full of inorganic minerals; eating no salt; drinking distilled water; eating a well balanced natural diet; proper exercise, breathing and personal hygiene; and the proper manner of positive thinking. Without a healthy body, one cannot do continuous positive thinking.

You now know that the most destructive force that can rob you of the feeling of youthfulness consists of a gradual precipitation of insoluble inorganic mineral matter from water and salt within the tissues of the body. The precipitation begins first in the walls of the arteries, gradually diminishing their elasticity and caliber and the nourishment of the tissues they supply. As a consequence of these destructive changes, all functional activities slow down more and more until some vital organ stops altogether, and death occurs from premature ageing.

The beginning of precipitation and the hardening of the arteries are the first stage of premature ageing, irrespective of the number of years a person has lived.

The choice of which road to take is up to the individual. He alone can decide whether he wants to reach a dead end or live a healthy, wholesome, long, active life.

So, your battle for life is a battle to keep the arteries free from inorganic minerals; keep chemicals and inorganic minerals from drinking water out of your body; stop the use of salt, and eat a balanced diet that does not leave a residue of toxic crystals to clog and obstruct the circulation.

You now know what I know — and I feel that I have the world's foremost health secret. Let's really make Life a healthful, vigorous adventure. Health and Happiness is our goal. Health is Wealth.

There are no failures for those who start in time, and who move steadily onward in quest of Perfect, Youthful Health.

Our sincere blessings to you dear friends, who make our lives so worthwhile and fulfilled by reading our teachings on natural living as our Creator laid down for us all to follow . . . Yes—he wants us all to follow the simple path of natural living and this is what we teach in our books and health crusades world-wide. Our prayers reach out to you for the best in health and happiness for you and your loved ones. This is the birthright He gives us all . . . but we must follow the laws He has laid down for us, so we can reap this precious health, physically, mentally and spiritually!

Patricia Bragg

"Teach me Thy way, O Lord;
and lead me in a plain path . . ."
Psalms 97:11

Many people go throughout life committing partial suicide — destroying their health, youth, beauty, talents, energies, creative qualities. Indeed, to learn how to be good to oneself is often more difficult than to learn how to be good to others.

Paul C. Bragg

Living is a continual lesson in problem solving, but the trick is to know where to start. No excuses—start your Health Program Today.

SPECIAL SUPPLEMENT

Since its first publication, "The Shocking Truth About Water" has been in such demand that the 23rd printing of the First Edition is on the bookstands as this is being written. It has, we feel, played a part in the development of widespread public awareness of environmental or ecological problems. Man is suddenly realizing that he cannot continue to contaminate this planet Earth and survive.

This Special Supplement to the Second Edition brings you important new information on this vital subject.

A NEW ERA OF PERSONAL ECOLOGY

The threat to our natural resources has become personalized in the threat to our health and our very lives. Technology has outstripped biology. The increasing mechanization and industrialization of our society — at first welcomed as a benefactor bringing creature comforts and labor-saving devices — are now revealed as a "Trojan Horse" bringing the enemy into our very homes to destroy us.

BIOCHEMISTS SHOCKED AT LAB TESTS

Biochemists are alarmed by the results of laboratory tests which reveal increasing deposits of inorganic "heavy metals" in our bodies. The dangerous effects of the increasing pollution of our air and water are evidenced by these figures:

- 90% of tests show mercury poisoning.
- 85% of tests show lead intoxication.
- 37% of tests show arsenic poisoning.
- 70% of tests show zinc accumulations.

POISONED BY FOOD AND WATER

Mercury . . . lead . . . arsenic . . . zinc . . . none of us deliberately take these inorganic mineral poisons into our systems. Or do we? How could we? Health-minded people drink water and fruit and vegetable juices, eat organic foods.

The tragic truth is that even these can become contaminated, because so much of our air and water and soil are contaminated by industrial and agricultural pollutants. We are aware of radioactive fallout and have demanded and been provided with safeguards against it. But what about the daily fallout of inorganic wastes from factory chimneys? What about poisonous pesticides, chemical fertilizers and food additives?

Rain water, for example, was once and rightly considered pure – but no longer. Pure it may be when it leaves the clouds. But after it passes through air polluted with industrial and automotive wastes – including everything from sooty carbons to strontium, arsenic, selenium, berylium, copper, lead, mercury and fluorides – it should be labeled "hazardous to your health".

When these poisons, especially the deadly fluoride gasses, are absorbed by the soil, it also becomes toxic. Inorganic minerals are taken from the soil by plants and, through the process of photosynthesis, are transformed into an organic state as food for animals, including man. This transformation from inorganic to organic multiplies by more than 500% the concentration of these toxic compounds. Add to these the contamination of poisonous pesticides and chemical fertilizers. Grains, vegetables and fruits can carry these poisons. So can meat, from animals that feed on contaminated grass and fodder. Is it small wonder that so many people are only half-alive?

AMINO ACIDS VITAL TO LIFE

From the germ cell of life to the most complicated living organism, amino acids are now known to be the activating ingredients of life itself. Amino acids are responsible for the production of proteins, the building blocks of the body, as well as the hormones and enzymes of thinking and memory, of breathing and muscle action.

Bio-Science has isolated some thirty distinct varieties of amino acids. Most of these are manufactured within our bodies, but some must be supplied daily from the food we eat. These amino acids are essential to life – life cannot "live" without them.

"HEAVY METALS" MURDER AMINO ACIDS

The discovery of amino acids was a tremendous scientific breakthrough. Now we are on the threshold of another — the unmasking of the murderer of these life-essentials. The substances which lead to the destruction of the body chemicals necessary in the metabolizing and manufacturing of these vital amino acids are proving to be — you guessed it! — inorganic "heavy metals" from industrial and agricultural contaminants.

"PLASTIC" FOOD STARVES OUT AMINO ACIDS

Trace organic minerals are also vital in the production of amino acids by the body. By natural law, these are supplied directly by fruit, vegetables and grains, and indirectly by dairy products and meats.

Today, however, we are being robbed of these essential elements. There are no trace organic minerals in the "plastic", demineralized, devitalized foods produced from depleted soils that are further deadened and contaminated by inorganic fertilizers and pesticides. The results are "pretty" but lifeless supermarket produce, and, secondly, poor quality livestock.

PROTECTIVE ACTION IN PERSONAL ECOLOGY

With an inorganic "heavy metal" toxicity caused by air and water pollutants, and a loss of the essential daily supply of trace minerals due to depleted and contaminated soil, it is not surprising that many recent tests are showing a dangerously low count of amino acids.

What can we do to protect ourselves?

These actions can be taken now, while we wait for more information on these important discoveries:

1. Learn more about amino acids and the enzymes that make them work, as well as the poisons that can kill them.

2. Have your physician make these two tests: a) for mercury, arsenic, lead, fluorides, strontium, and all "heavy metals"; b) the comparatively new test known as the "30 amino acids fractionation test".

3. Encourage your friends and loved ones to seek the assistance of their doctors in measuring their bodies for these malfunctions before they become diseases.

4. Make sure that the source of your organic food supply is in an area free from airborne pollution, and produced in soil enriched by natural fertilizers.

5. Follow the precepts of this book in regard to your drinking water! Drink only steam-distilled water.

POLITICAL ACTION FOR PERMANENT ECOLOGY

During recent years — in response to an aroused public opinion — political platforms and the politicians who stand on them, from local to national, have promised action in solving our ecological problems. Millions of dollars have been spent on "studies", volumes of words printed and spoken.

But the ever increasing problems of pollution and a dangerously disrupted ecology are still with us. The air in our cities is fast approaching crisis conditions. Our drinking water sources, our rivers, lakes and ocean shorelines are becoming contaminated far beyond safe levels.

Scientific bodies throughout the world agree that we are in danger of self-extinction, unless we stop abusing our vital life-support systems. They have offered sober warnings, with countdown time tables. So far, these warnings have received but token hearings from political ears.

But there is one voice that every politician heeds — the concerted voice of aroused voters who demand action. This is the one voice that rises above the clink of campaign contributions from big industrial, agricultural, manufacturing, financial and business interests.

To achieve the desired effect — political action on any and all levels — the voice of the voters must be strong, widespread, and united in purpose.

WHAT CAN ONE PERSON DO?

Whether you are counting to ten or to a million, you have to begin with one. That's how political action starts. Begin with yourself, your family, your friends and neighbors . . . involve your club, your church, other groups. Get your local newspaper into the act by writing letters to the editor. Call and write your local radio and TV stations. Attend town council meetings and speak your piece.

100

After you get your hometown or community involved, extend your action into the county . . . the state . . . the nation. Write and call on county officials, state officials and legislators, your congressman and senator.

Don't settle for promises or palliatives. Demand action — and persist until you get it.

It is up to each and every one of us to bring all possible political pressure to bear upon our political leaders to clean up our environment and restore natural ecology. This is what we must do — if we hope to see the 21st Century.

LET'S CLEAR UP CONFUSION

On the basic issues of pollution and ecology, practically all health minded persons agree.

On other issues among ourselves, however, there seems to be some confusion. On the subject of nutrition, for example, there was so much controversy among the advocates of various diets -- each claiming his/hers to be the only one worthwhile -- that I felt it necessary to clarify matters. As the world's oldest practicing biochemist and nutritionist, with personal experience in evaluating the numerous and varied dietetic theories, I have shared this knowledge with all those who are interested in our revealing book, **"Healthful Eating Without Confusion."** See back pages for ordering this book.

Now I find a similar controversy developing in regard to water.

WHAT IS "PURE WATER"?

Having read this book on "The Shocking Truth About Water", you have become aware of the pros and cons of Distilled Water versus Mineral or Ground Water.

Unfortunately, however, writers and lecturers are now creating confusion about Distilled Water. Some have actually referred to "soft water" (treated through a water softener) as "distilled water". This is definitely not the case. Soft water has a high content of sodium, calcium, and other inorganic minerals.

Using this misinformation as a basis, even some health publications have made fallacious comparisons in an attempt to make a case against Distilled Water! For example, group comparisons were made citing people with a low incidence of cardiovascular (heart and circulatory system) problems, who lived in remote areas with hard ground water supplies — without taking into consideration also the pertinent fact that such people eat more natural "live" foods and live in a relaxed rural environment.

On the basis of water alone, this rural group was compared with people living in crowded, highly competitive metropolitan areas consuming treated city water, presumably filtered through home water softeners (erroneously designated as "distilled water"), who showed a higher incidence of hypertension and other cardiovascular problems. No mention was made of the obvious facts that these urban dwellers were subjected to much greater tensions, as well as a diet of lifeless, processed "plastic" foods from supermarket shelves — major contributors to cardiovascular illnesses.

Obviously, any self-styled "researcher" who makes such errors as the above has not done his/her basic homework! Unfortunately, however, the reader may not be aware of this — and thus unnecessary confusion is irresponsibly created.

So, with water as with eating, don't let the "experts" fool you!

WHAT IS DISTILLED WATER?

In this book, I have stressed that Distilled Water is the only pure water — the only water you should put into your body.

As noted previously, "soft water" is <u>not</u> Distilled Water. Neither is "purified water" . . . nor "ionized water" . . . nor "reverse osmosis water".

There is only one process that can make 100% Distilled Water, and that is Steam Distillation. In Steam Distillation, only pure water, H_2O, evaporates, leaving all inorganic minerals and other impurities behind.

I have researched the processes and equipment available for making Distilled Water, and continue to review and experiment with new equipment as it comes on the market. Information as to the best home steam water distillers on the market, in my estimation, is available free of charge from

Pure Water Systems Research
7340 Hollister Ave., Goleta, California 93117

COMMENTS ON
"THE SHOCKING TRUTH ABOUT WATER"

In my opinion, "The Shocking Truth About Water" is destined to become a landmark — not just in the field of so-called "health literature" — but particularly in the field of standard internal medicine.

Following a massive coronary thrombosis 13 years ago, I have been on a strict regimen of distilled water, health diet, vitamin therapy and exercise — and the results have exceeded my expectations. I feel better at 67 than I did at 47. My arteries have softened; my joints have limbered; my vision is sharper; my nerves are calmer; and my head is clearer.

My own experience corroborates your findings. I am convinced that distilled water has been the most important facet of my rejuvenation program.

— Ben H. Martin, Lakewood, Calif.

Thank you for your fine review of my books, "Hunza Land" and "The Choice Is Clear", in a recent issue of your excellent HEALTH BUILDER. And let me repeat my congratulations on your book, "The Shocking Truth About Water", which I recommend as a "must" to all my patients.

In connection with this subject, I would like to emphasize that a vital factor in the amazing longevity of the people of Hunza Land is distilled water. They eat most of their fruits and vegetables raw, raised in organic soil. Fruits and vegetables, of course, are 90% distilled water — nature's own distillation, as you say. Along with that, they drink glacier water, which is low in inorganic minerals. Wine is their main beverage, which again is distilled water. So, in isolated Hunza Land, the intake of distilled water is 90% greater than that of our vaunted western civilization.

— Allen E. Banik, O.D., Kearney, Nebr.

I have read your book, "The Shocking Truth About Water", and find it indeed shocking. Thank you for the enlightenment. This should be required knowledge in every medical school and health related field.

— Chris R. Linville, M.D., New Brunswick, N.J.

For the past few years I have been doing research work on hardening of the arteries and related problems of aging. Your book, "The Shocking Truth About Water", is the best and most reasonable I've read on this subject.

— Betty Watts, Pasadena, Calif.

I belong to a Health Club in Buffalo, N.Y., and our members all agree that your wonderful book, "The Shocking Truth About Water", is the best book on the market.

— H.W. Hoffman, Hamburg, N.Y.

One of the greatest things that ever happened to me was attending your health classes 35 years ago in Miami, Florida. Thanks to your teachings, I am now 57 years young and feel better than ever . . . I have read all of your books, some of them many times, and have just finished the sixth reading of "The Shocking Truth About Water." To me, this is the greatest book ever published. Keep up the good work!

— Cliff Hayes, Deerfield Beach, Fla.

I have been on distilled water for about five months, ever since I read your book. I had pains in my knuckles and a calcium deposit on my left shoulder. These have all left me now; also my bowel elimination is a lot better.

— C.A. McFeaters, Hainesville, Penn.

As outlined in your book, "The Shocking Truth About Water," I know that the closer to nature we can get, the better off we are going to be . . . I am a farmer in Missouri, and have not used any chemicals of any kind on this farm for 15 years, as I came to realize that we should not try to improve on nature but work with her . . . I am so happy to find people like you, who are not trying to keep the truth to yourselves and are leading people toward health.

— Eugene Kling, Meadville, Mo.

From start to finish, this is the most enthralling book I've had the pleasure of reading.

— Mrs. Elizabeth Fisch, Ridgefield, N.J.

Your book on water is a masterpiece!

— H.W. Rosenthal, Richmond Hill, Ont., Canada

QUESTIONS AND ANSWERS

Q. *"The Shocking Truth About Water"* *is so informative that I wish to give copies of it to my children and families and friends. Do you have a special price on a dozen copies?*
A. Yes, there is a special rate on the purchase of a dozen or more copies of any BRAGG publication. In fact, many church groups, service clubs and school organizations buy BRAGG Books in quantity lots for individual resale at regular prices as a fund-raising project.

Q. *Will distilled water help my complexion?*
A. Distilled water will help you have a smooth, firm, radiant complexion in two ways — by drinking it for internal cleanliness, and by cleansing your skin with it externally. Hard water seals the pores and tends to clog them. For thorough, healthful cleansing, use distilled water and a pure soap (from your Health Food Store) on your skin. Distilled water is also excellent as a hair rinse.

Q. *I have an obesity problem. Will drinking distilled water help me to lose weight?*
A. First, eliminate salt from your diet. A major cause of obesity is retention of fluid in the tissues, and a major cause of water-logged tissues is the fact that salt is indigestible by the human body and must be held in solution with water. Hard water makes this condition worse by adding more indigestible inorganic minerals which impair the body's system of elimination. Pure, distilled water will help your body to function at its best in every way, including the elimination of accumulated harmful inorganic substances. Distilled water is especially kind to your liver and kidneys, which are the organs most abused by salt and hard water.

Q. *Is distilled water recommended for babies?*
A. It is not only recommended — it is prescribed for them by doctors. Distilled water should be used not only internally but also externally for babies, as diaper rash and other skin problems can result from hard water deposits even on freshly washed clothing.

105

Q. *What is your opinion of the honey and vinegar treatment for arthritis advocated by Dr. Jarvis?*

A. It is an excellent treatment — except that he should have specified distilled water. A quarter cup of raw apple cider to a gallon of distilled water, flavored with raw honey, makes a drink "fit for the gods" — and will help you to feel like one. But remember, it took years to build up your arthritic condition, so don't expect overnight "miracles". Work with mother Nature and be as patient as she is.

Q. *Can animals and wild life tell the difference in water?*

A. Yes. Place the 9 different kinds of water before a goat, for example, and he will select the distilled water. Keep distilled water in your bird bath, and the same birds will return year after year. Many a horse race has been lost because trainers did not provide thoroughbreds with distilled water.

Q. *Does hard water affect everyone the same way?*

A. No. Although all human systems are basically similar, no two are exactly alike. Mineral deposits from hard water tend to seek the weakest points, producing varied symptoms — on the intestinal walls, constipation; in the kidneys, stones; in the arteries, artereosclerosis; in the joints, arthritis; and so on. Of course, when the functions of any one part of the body are impaired, the entire system is affected and gradually weakened — thus multiple symptoms appear as evidence of more widespread damage. Your best insurance against all these "aging" symptoms is to drink distilled water.

Q. *Do athletes drink distilled water?*

A. The wise ones do. The famous Connie Mack, manager of the Yankees for some 30 years, would not allow his players to drink hard water on any occasion — and he had wonderfully healthy teams. Connie Mack also "practiced what he preached" and maintained his own perfect health to age 90.

Q. *We have recently installed a home water softener. How does this affect the water for drinking purposes?*

A. Don't drink it! Water softeners do not eliminate the inorganic minerals, but merely hold them in suspension in an ionized state. This will make more soapsuds — but will leave the same mineral deposits in both home and human plumbing!

UPDATE ON DRINKING WATER
AND
HOW IT EFFECTS YOUR HEALTH

Pure water is a necessity for health!

In research begun way back in 1960, the water supplies of 1,633 of the largest cities in the U.S. were analyzed. Results of this long study showed a definite link between water quality and the mortality rate from cardiovascular, carcinogenic and other chronic diseases.

TDS AND CHRONIC DISEASE

In his study, "Relationship of Water to the Risk of Dying," Dr. Sauer chronicled the relationship of TDS (Total Dissolved Solids) to heart disease, cancer and other chronic diseases. Total Dissolved Solids is the term for all the elements present in any water supply. It had been thought for centuries that the European mineral waters so very high in total dissolved solids were beneficial to health. But as TDS increases in a water supply, so does the number of chronic diseases increase in the population using that supply.

(Author's Note: At our home in Desert Hot Springs, one of California's renowned hot mineral water resorts, our water pipes had to be replaced after only a few years due to mineral buildup in the pipes. What water quality does to plumbing in building, it also does to human pipes!)

HIGH BLOOD PRESSURE AND DRINKING WATER

Water quality also plays a part in hypertension, or high blood pressure. Hypertension afflicts over 24 million in the U.S.A. making it the most common chronic disease. It is a major health problem also in all of the developed countries of the world, due to stress, refined foods, salt, lack of exercise, etc. But the good news about hypertension, or high blood pressure, is that it can be reduced or even prevented! A diet that is high in natural fiber, grains, vegetables, calcium and potassium—with less meat, fat and sodium, can effectively help reduce hypertension.

It's estimated that fully 10% of our sodium intake is from drinking water. A study of high school sophomores in a community with high levels of sodium in the drinking water, showed significantly higher blood pressure levels than in areas with less sodium in the water supply. The girls among this first group had blood pressure patterns characteristic of persons 10 years older. Follow up studies among even younger children, ages 7 to 11 years, in the same area produced similar results.

The conclusion from this and other studies seems obvious: increased sodium levels in drinking water result in increased blood pressure levels. The American Heart Association, the EPA (Environmental Protection Agency) and the World Health Organization—among other health groups—recommend that sodium levels in drinking water should not exceed 20 mg/liter. And yet, of 2,100 water supplies surveyed by the U.S. Public Health Service, 42% had sodium ion concentrations above this level. About 5% showed levels greater than 250 mg/liter.

WATER SOFTENERS AND SODIUM EQUALS TROUBLE

We have already seen the relationship between soft water and heart disease! But that is only part of the story—for softened water poses grave health risks in terms of hypertension also.

The usual method for softening water is to add two parts sodium which then extracts one part calcium and one part magnesium from the water supply. This results in water low in hardness but higher in sodium.

So it would seen that the "luxury" of having soft water to bathe in and for laundry—is hardly worth the increased risks to health.

THE SAD TRUTH ABOUT CHLORINATION

Water chlorination has been widely used to "purify" water in this country for most of this century. But surely its negative effects on health outweigh any benefits.

Dr. Joseph Price, for one, believes there is a definite link between widespread chlorination of water supplies and the increasing incidence of heart disease! In animal experiments he conducted, chlorine caused atherosclerosis in 95% of the animals tested!

Chlorine in the water supply has been correlated with cancers of the bladder, liver, pancreas and urinary tract in certain areas. In New Orleans, for just one example of what is happening around the world—drinking water is taken from the Mississippi River. There, over 66 new carcinogenic compounds have been isolated in the water supply as a result of adding chlorine that combines with methanol, carbon disulfide and other compounds. A very high incidence of colon cancer is associated with this area.

One expert, Dr. Herbert Schwartz, is quite emphatic in asserting that "chlorine is so dangerous it should be banned." He believes that chlorine-treated water is directly responsible for cancer, heart disease and premature senility!

CANCERS AND CHLORINATION

An investigator at the government's National Cancer Institute, Kenneth Cantor, points out that many studies since the 1974 report on New Orleans have confirmed its findings, linking increased carcinogens in the water supply to additional cancer deaths annually. Cantor himself , with associates, completed a study of nearly 3,000 men and women who have been drinking chlorinated water in such cities as New York, Chicago, Atlanta, Detroit, New Orleans, San Francisco and Seattle. Subjects were also studied in Connecticut, Iowa, New Jersey, New Mexico and Utah. His study conclusively linked bladder cancer and chlorinated drinking water.

Nor is the risk limited to bladder cancer. Theresa Young of the Department of Preventive Medicine at the University of Wisconsin led a study to determine the effect of chlorinated water on women. She checked the death certificates of women in Wisconsin who had died from cancers of the gastrointestinal system, the urinary tract, brain, lungs and breast.

The major finding was that colon cancer in women was "significantly associated" with exposure to water that was disinfected with low, medium and high daily chlorine doses for a least 20 years. He said her study should be examined in the context of the theory linking colon cancer to a high fat, low fiber diet. She believes researchers should pursue other theories of colon cancer as well—such as industrial pollutants and chlorine-induced carcinogens in drinking water.

MISCARRIAGES, BIRTH DEFECTS, ? IN TAP WATER

A recent $1 million, 5,000 woman study by the State of California revealed women who drank tap water had twice as many miscarriages, children with birth defects as women who drank bottled water or had filtering devices on their taps. Five large studies have come to the same conclusion, according to State Health Director Kenneth Kizer. Only two aspects of health were investigated. It is impossible even to guess how many other human ailments for men, women and children can be contributed to by tap water.

DANGEROUS CHEMICALS IN OUR DRINKING WATER

There is more and more evidence that the majority of human cancers are environmental in origin and thereby largely preventable. In fact, an astounding number of chemical—and possibly carcinogenic compounds are to be found in our water, after treatment as well as in surface and ground water resources.

A recent ABC News expose revealed the shocking fact that over 700 chemicals have already been found in our drinking water. Of these, 129 have been pinpointed by the EPA as posing serious health risks. Yet the agency requires that our water supplies be tested for only 14 of them.

One carcinogen found in many municipal water systems, chloroform, can be introduced during chlorine treatment. A known animal carcinogen, it is present in measurable levels of nearly all municipal water systems as a by-product of water chlorination!

FLOURIDE AND CANCER

Fluoride is among the most potentially dangerous of all water additives. Long term research into fluoridation has shown that its positive effects on dental health are minimal at best and are far outweighed by the serious health risks resulting from its use.

110

Cancer researcher Dr. Dean Burk believes that 10% of all cancer deaths in the U.S. may be directly due to fluoridation. Yet 40% of the citizenry continues drinking fluoridated water. Dr. Burk and his associate Dr. John Yiamouyiannis, have concluded that drinking artificially fluoridated water may increase your risk of dying from cancer by 5 to 15%.

DRINKING SAFE WATER

With so many of our public water supplies contaminated by chemicals and artificial additives, how can we make sure that we provide only safe, healthful water for ourselves and our families?

One relatively simple answer is to drink only distilled, bottled water. But even this method requires care, for not all bottled water lives up to even FDA Bottled Water Standards. Before you buy any bottled water, ask for an analysis—it's your right—it's your health at risk! Let's protect these water atrocities and get quick action to clean up U.S.A. water supplies!!! Call or write to the bottled water company and also you city water board, to be sure it contains no THMs (trihalomethanes also formed through chlorination), carcinogens, synthetic organic chemicals or pesticides!

HOME WATER DISTILLING & FILTTRATION SYSTEMS

Another method of ensuring a safe water supply is to install a home water distilling or filtration system (change filters often) to purify your own tap water. But, as with bottled water, any filtration system must be evaluated and its lab report given careful study. Some systems remove up to 99% of THMs and synthetic organic compounds. But the main value of less efficient systems is to provide a false sense of security.

PURE WATER: THE GREAT LIFE GIVER!

Pure water (distilled is best) is truly one of God's greatest gifts to us, a source of life and health. And making sure that we use only water that is safe and uncontaminated may be one of the greatest health gifts we can give ourselves!

WATER REVIEW AND UPDATE

Distilled water is one of the world's best and purest waters! It is excellent for detoxification and fasting programs and for helping clean out all the cells, organs, and fluids of the body because it can help carry away so many harmful substances!

Water from chemically-treated public water systems and even from many wells and springs is likely to be loaded with poisonous chemicals and toxic trace elements.

Depending upon the kind of piping that the water has been run through, the water in our homes, offices, schools, hospitals, etc., is likely to be overloaded with zinc (from old-fashioned galvanized pipes) or with copper and cadmium (from copper pipes). These trace elements are released in excessive quantity by the chemical action of the water on the metals of the water pipes.

PURE WATER — ESSENTIAL FOR HEALTH!

Yes, pure water is essential for health, either from the natural juices of vegetables, fruits, and other foods, or from the water of high purity obtained by steam distillation which is the best method, or by one of the new high-efficiency deionization processes.

The body is constantly working for you . . . breaking down old bone and tissue cells and replacing them with new ones. As the body casts off the old minerals and other products of broken-down cells, it must obtain new supplies of the essential elements for the new cells. Moreover, Scientists are only now beginning to understand that various kinds of dental problems, different types of arthritis, and even some forms of hardening of the arteries are due to varying kinds of imbalances in the levels of calcium, phosphorus, and magnesium in the body. Disorders can also be caused by imbalances in the ratios of various minerals to each other.

Each individual healthy body requires a proper balance within itself of all the nutritive elements. It is just as bad for any individual to have too much of one item as it is to have too little of that one or of another one. It takes appropriate levels of phosphorus and magnesium to keep calcium in solution so it can be formed into new cells of bone and teeth. Yet, there must not be too much of those nor too little calcium in the diet, or old bone will be taken away but new bone will not be formed.

In addition, we now know that diets which are unbalanced and inappropriate for a given individual can deplete the body of calcium, magnesium, potassium, and other major and minor elements.

Diets which are high in meats, fish, eggs, grains, nuts, seeds, or their products may provide unbalanced excesses of phosphorus which will deplete calcium and magnesium from the bones and tissues of the body and cause them to be lost in the urine.

A diet high in fats will tend to increase the uptake of phosphorus from the intestines relative to calcium and other basic minerals. Such a high-fat diet can produce losses of calcium, magnesium, and other basic minerals in the same way a high-phosphorus diet does.

Diets excessively high in fruits or their juices may provide unbalanced excesses of potassium in the body, and calcium and magnesium will again be lost from the body through the urine.

Deficiencies of calcium and magnesium . . . for example can produce all kinds of problems in the body ranging from dental decay and osteoporosis to muscular cramping, hyper-activity, muscular twitching, poor sleep patterns, and excessive frequency of uncontrolled patterns of urination. Similarly, deficiencies of other minerals, or imbalances in the levels of those minerals, can produce many other problems in the body.

Therefore, it is important to clean and detoxify the body through fasting and through using distilled or other pure water as well as healthy organically-grown vegetable and fruit juices. At the same time, it is also important to provide the body with adequate sources of new minerals. This can be done by eating a widely-distributed diet of wholesome vegetables, including kelp and other sea vegetables for adults and healthy mother's milk for infants, and certified raw goat's or cow's milk for those children and adults who are not adversely affected by milk products . . . but most processed home homogenized milks we do not suggest using.

But, despite dietary sources such as these, many adults and children in so-called civilized cultures will be found to have low levels of essential minerals in their bodies due to losses caused by coffee, tea, carbonated beverages, and long-term bad diets containing too much sugar and other sweets as well as products made from refined flours and containing refined table salt.

In addition,the body's organ systems can be thrown out of balance by continuing stress, by toxins in our air and water, by disease-produced injuries, and by pre-natal deficiencies in the mother's diet or life style.

As a result, many, if not most people in our so-called civilization may need to take mineral supplements such as the new chelated multiple mineral preparations as well as a broad-range multiple-vitamin tablet.

COMMON SENSE REASONS
WHY YOU SHOULD DRINK PURE,
DISTILLED WATER!

- There are over 12 thousand chemicals on the market today, 500 being added yearly. Regardless of where you live, in the city or on the farm, some of these chemicals could be getting into your drinking water.

- No one on the face of the earth today knows what effect these chemicals could have, as they go into thousands of different combinations. It is like making a mixture of colors; one drop could change the complete color.

- There has not been equipment designed to detect some of these chemicals, and there may not be for many years to come.

- The body is made up of approximately 65% water. Therefore, don't you think you should be particular about the type of water you drink?

- The Navy has been drinking distilled water for several generations!

- Distilled water is chemical and mineral free. Distillation removes the chemicals and impurities from water that are possible to remove, and if distillation doesn't remove them, there is no known method today that will.

- The body does need minerals . . . but it is not necessary that they come from water. There is not one mineral in water which cannot be found more abundantly in food! Water would be a most unreliable source of minerals because it varies from one area to another. The food we eat, not the water we drink, is our most reliable source of minerals!

- Distilled water is used for intravenous feeding, inhalation therapy, prescriptions and baby formulas. Therefore, doesn't it make common sense that distilled water is good for everyone?

- Thousands of distillers have been sold throughout the United States and in many foreign countries, to individuals, families, dentists, doctors, hospitals, nursing homes and government agencies . . . and these informed alert consumers are helping protect their health by using pure distilled water.

- With all of the chemicals, pollutants and other impurities in our water, doesn't it only make good common sense that you should clean up the water you drink, the inexpensive way, through distillation — nature's way of purifying water.

FROM THE AUTHORS

This book was written for YOU. It can be your passport to the Good Life. We Professional Nutritionists join hands in one common objective — a high standard of health for all and many added years to your life. Scientific Nutrition points the way — Nature's Way — the only lasting way to build a body free of degenerative diseases and premature aging. This book teaches you how to work with Nature and not against her. Doctors, dentists, and others who care for the sick, try to repair depleted tissues which too often mend poorly if at all. Many of them praise the spreading of this new scientific message of natural foods and methods for long-lasting health and youthfulness at any age. To speed the spreading of this tremendous message, this book was written.

Statements in this book are recitals of scientific findings, known facts of physiology, biological therapeutics, and reference to ancient writings as they are found. Paul C. Bragg has been practicing the natural methods of living for over 70 years, with highly beneficial results, knowing they are safe and of great value to others, and his daughter Patricia Bragg works with him to carry on the Health Crusade. They make no claims as to what the methods cited in this book will do for one in any given situation, and assume no obligation because of opinions expressed.

No cure for disease is offered in this book. No foods or diets are offered for the treatment or cure of any specific ailment. Nor is it intended as, or to be used as, literature for any food product. Paul C. Bragg and Patricia Bragg express their opinions solely as Public Health Educators, Professional Nutritionists and Teachers.

Certain persons considered experts may disagree with one or more statements in this book, as the same relate to various nutritional recommendations. However, any such statements are considered, nevertheless, to be factual, as based upon long-time experience of Paul C. Bragg and Patricia Bragg in the field of human health.

TOTAL HEALTH FOR THE TOTAL PERSON

In a broad sense, "Total Health for the Total Person" is a combination of physical, mental, emotional, social, and spiritual components. The ability of the individual to function effectively in his environment depends on how smoothly these components function as a whole. Of all the qualities that comprise an integrated personality, a well-developed, totally fit body is one of the most desirable.

A person may be said to be totally physically fit if they function as a total personality with efficiency and without pain or discomfort of any kind. That is to have a Painless, Tireless, Ageless body, possessing sufficient muscular strength and endurance to maintain an effective posture, successfully carries on the duties imposed by the environment, meets emergencies satisfactorily and has enough energy for recreation and social obligations after the "work day" has ended, meets the requirements for his environment through efficient functioning of his sense organs, possesses the resilience to recover rapidly from fatigue, tension, stress and strain without the aid of stimulants, and enjoys natural sleep at night and feels fit and alert in the morning for the job ahead.

Keeping the body totally fit and functional is no job for the uninformed or the careless person. It requires an understanding of the body, sound health and eating practices, and disciplined living. The results of such a regimen can be measured in happiness, radiant health, agelessness, peace of mind, in the joy of living and high achievement.

Paul C. Bragg and Patricia Bragg

"I have found a perfect health, a new state of existence, a feeling of purity and happiness, something unknown to humans . . ."
—Novelist Upton Sinclair,
who fasted frequently.

"Give us, Lord, a bit of sun,
A bit of work and a bit of fun.
Give us, in all struggle and sputter,
Our daily whole grain bread and a bit of nut butter.
Give us health, our keep to make
And a bit to spare for others' sake.
Give us, too, a bit of song
And a tale and a book, to help us along.
Give us, Lord, a chance to be
Our goodly best for ourselves and others
'Til all men learn to live as brothers."

— An Old English Prayer

The preservation of health is a duty.
Few seem conscious that there is
such a thing as physical morality.

To my mind the greatest mistake a person can make is to remain ignorant when he is surrounded, every day of his life, by the knowledge he needs to grow and be healthy and successful. It's all there. We need only to observe, read, learn . . . and apply.

Your birthday is the beginning of your own personal new year. Your first birthday was a beginning, and each new birthday is a chance to begin again, to start over, to take a new grip on life.

WE THANK THEE

For flowers that bloom about our feet;
    For song of bird and hum of bee;
For all things fair we hear or see,
    Father in heaven we thank Thee!
For blue of stream and blue of sky;
    For pleasant shade of branches high;
For fragrant air and cooling breeze;
    For beauty of the blooming trees,
Father in heaven, we thank Thee!
    For mother-love and father-care,
For brothers strong and sisters fair;
    For love at home and here each day;
For guidance lest we go astray,
    Father in heaven, we thank Thee!
For this new morning with its light;
    For rest and shelter of the night;
For health and food, for love and friends;
    For every thing His goodness sends,
Father in heaven, we thank Thee!

— *Ralph Waldo Emerson*

Hard Water Hurts Skin

According to the British Medical Journal, hard water may cause hands to become dry and sore. It is the mineral content of the water, particularly calcium hardness compounds, that acts as irritant. Because more soap must be used with hard water to obtain a lather, this can be a further source of irritation.

Take time
for **12** things

1 Take time to Work—
it is the price of success.

2 Take time to Think—
it is the source of power.

3 Take time to Play—
it is the secret of youth.

4 Take time to Read—
it is the foundation of knowledge.

5 Take time to Worship—
it is the highway of reverance and washes the
dust of earth from our eyes.

6 Take time to Help and Enjoy Friends—
it is the source of happiness.

7 Take time to Love—
it is the one sacrament of life.

8 Take time to Dream—
it hitches the soul to the stars.

9 Take time to Laugh—
it is the singing that helps with life's loads.

10 Take time for Beauty—
it is everywhere in nature.

11 Take time for Health—
it is the true wealth and treasure of life.

12 Take time to Plan—
it is the secret of being able to have time to
take time for the first eleven things.

God grant me
the serenity
to accept the things
I cannot change —
the courage
to change the things
I can —
And the wisdom to know
the difference —

Anonymous

"Why not look for the best — the best in others, the best in ourselves, the best in all life situations? He who looks for the best knows the worst is there but refuses to be discouraged by it. Though temporarily defeated, dismayed, he smiles and tries again. If you look for the best, life will become pleasant for you and everyone around you."

— Paul S. Osumi

FOOD FOR THOUGHT

Soup rejoices the stomach, and disposes it to receive and digest other food.
— Brillat Savarin

To work the head, temperance must be carried into the diet. — Beecher

To fare well implies the partaking of such food as does not disagree with body or mind. Hence only those fare well who live temperately. — Socrates

The eating of much flesh fills us with a multitude of evil diseases and multitudes of evil desires. — Porphyrises, 233 A.D.

Health is not quoted in the markets because it is without price.

It is a mistake to think that the more a man eats, the fatter and stronger he will become.

The health journals and the doctors all agree that the best and most wholesome part of the New England country doughnut is the hole. The larger the hole, they say, the better the doughnut.

According to the ancient Hindu Scriptures, the proper amount of food is half of what can be conveniently eaten.

The nervousness and peevishness of our times are chiefly attributable to tea and coffee. The digestive organs of confirmed coffee drinkers are in a state of chronic derangement which reacts on the brain, producing fretful and lachrymose moods. — Dr. Bock, 1910

A physician recommended a lady to abandon the use of tea and coffee. "O, but I shall miss it so," said she. "Very likely," replied her medical adviser, "but you are missing health now, and will lose it altogether if you do not."

WATER

To the days of the aged it addeth length;
To the might of the strong it addeth strength;
It freshens the heart, it brightens the sight;
'Tis like quaffing a goblet of morning light.

PATRICIA BRAGG, N.D., Ph.D.

Angel of Health

Lecturer, Author, Nutritionist, Health Educator & Fitness Advisor to World Leaders, Glamorous Hollywood Stars, Singers, Dancers & Athletes.

Daughter of the world renowned health authority, Paul C. Bragg, Patricia Bragg has won international fame on her own in this field. She conducts Health and Fitness Seminars for Women's, Men's, Youth and Church Groups throughout the world...and promotes Bragg "How-To, Self-Health" Books in Lectures, on Radio and Television Talk Shows throughout the English-speaking world. Consultants to Presidents and Royalty, to Stars of Stage, Screen and TV, and to Champion Athletes, Patricia Bragg and her father are Co-Authors of the Bragg Health Library of Instructive, Inspiring Books.

Patricia Bragg herself is the symbol of perpetual youth and super energy. She is a living and sparkling example of hers and her father's healthy lifestyle precepts.

A fifth generation Californian on her mother's side, Patricia Bragg was reared by the Natural Health Method from infancy. In school, she not only excelled in athletics but also won high honors in her studies and her counseling. She is an accomplished musician and dancer...as well as tennis player, swimmer and mountain climber...and the youngest woman to ever be granted a U.S. Patent. Patricia Bragg is a popular gifted Health Teacher and a dynamic, in-demand Talk Show Guest where she spreads simple, easy-to-follow health teachings for everyone.

Man's body is the Temple of the Holy Spirit, and our Creator wants us filled with Joy and Health for a long walk with Him for Eternity. The Bragg Crusade of Health and Fitness (3 John 2) has carried her around the world...spreading physical, spiritual, emotional and mental health and joy. Health is our birthright, and Patricia teaches how to prevent the destruction of our health from man-made wrong habits of living.

Patricia's been Health Consultant to American Presidents and to the British Royal Family, to Betty Cuthbert, Australlia's "Golden Girl" who holds 16 world records and four Olympic gold medals in women's track and to New Zealand's Olympic Track Star Allison Roe. Among those who come to her for advice are some of Hollywood's top stars from Clint Eastwood to the ever youthful singing group The Beach Boys and their families, singing stars of the Metropolitan Opera and top ballet stars. Patricia's message is of world-wide appeal to people of all ages, nationalities and walks-of-life who read the Bragg Health Books and attend the Crusades. ◆

Jesus said: "Thy faith hath made thee whole, and go and sin no more." And that means your dietetic sins. He Himself, through fasting and prayer, was able to heal the sick and cure all manner of diseases.

Dear friend, I wish above all things that thou may prosper and be in health even as the soul prosper. 3 John 2

BRAGG "HOW-TO, SELF-HEALTH" BOOKS
Authored by America's First Family of Health
Live Longer — Healthier — Stronger Self-Improvement Library

✔	BRAGG BOOK TITLES ORDER FORM	PRICE	Amts	$ TOTAL
__	Vegetarian Gourmet Health Recipes (No Salt, No Sugar, Yet Delicious)	$6.95		
__	Bragg's Complete Health Gourmet Recipes for vital health—448 pages	7.95		
__	The Miracle of Fasting (Bragg Bible of Health for Physical Rejuvenation)	6.95		
__	*Bragg Health & Fitness Manual for All Ages—Swim-Bike-Run A Must for Athletes, Triathletes & Would-Be Athletes - 600 pgs.	16.95		
__	Build Powerful Nerve Force (Reduce Stress, Fear, Anger, Worry)	5.95		
__	Keep Your Heart & Cardio-Vascular System Healthy & Fit at Any Age	5.95		
__	The Golden Keys to Internal Physical Fitness...	2.95		
__	The Natural Way to Reduce	5.95		
__	The Shocking Truth About Water (Learn Safest Water to Drink & Why)...................	5.95		
__	Your Health and Your Hair, Nature's Way to Beautiful Hair (Easy-To-Do Method) ..	5.95		
__	Healthful Eating Without Confusion (Removes Doubt and Questions)	4.95		
__	Salt-Free Sauerkraut Recipes (Use Our Delicious Recipes or Learn To Make Your Own) ...	2.95		
__	Nature's Healing System to Improve Eyesight in 90 days (Foods, Exercises, etc) ...	5.95		
__	Building Strong Feet	4.95		
__	Super Brain Breathing for More Energy & Vital Living (Can Double Your Energy) .	3.95		
__	Toxicless Diet-Purification & Healing System (Stay Ageless Program)	3.95		
__	Powerful Health Uses of Apple Cider Vinegar (How to Live Active to 120)..............	3.95		
__	Fitness/Spine Motion — For More Flexible, Painfree Back	3.95		
__	Building Health & Youthfulness	1.75		
__	Natural Method of Physical Culture	1.75		
__	Nature's Way to Health — Live 100 Active Years	3.95		
__	The Philosophy of Super Health	1.75		
__	South Sea Abdomen Culture for Perfect Elimination & Trim Waist	1.75		

HEALTH SCIENCE ISBN Prefix 0-87790

TOTALS	
Calif. Residents add 6% Tax	
Shipping	
Total Amount Enclosed $	

Shipping: **Please Add $1.50 for 1st book,**
and 75¢ for each additl. book,
or $3 each for airmail.
Add $3. for each Fitness Manual.

retail book orders over $35, postage is free. US funds only. Prices subject to change without notice.

lease Specify: ☐ Check ☐ Money Order ☐ Cash ☐ Credit Card

harge My Order To: ☐ Visa ☐ MasterCard

[MasterCard]   Card Expires: [MO.] [YR.]   MasterCard InterBank No. [ ][ ][ ][ ] (No. above name on card)

[VISA]   Credit Card Number: [ ][ ][ ][ ] [ ][ ][ ][ ] [ ][ ][ ][ ] [ ][ ][ ][ ]

Signature: _____

CREDIT CARD CUSTOMERS USE OUR FAST PHONE SERVICE: (805) 968-1020   In a hurry? Call (805) 968-1020. We can accept MasterCard or VISA phone orders only. Please prepare your order using this order form. It will speed your call and serve as your order record. Hours: 9am to 4pm Pacific Time, Monday to Thursday. FAX (805) 968-1001.

ail to: HEALTH SCIENCE, Box 7, Santa Barbara, CA 93102 U.S.A.
lease Print or Type)   (Be sure to give street and house number to facilitate delivery)
ROF-8901

me _____

dress _____   Apt. No. _____

y _____   State _____

) _____   Zip [ ][ ][ ][ ][ ]

 one

rder your Bragg Health Books Today...For a Healthier Tomorrow!

SEND FOR IMPORTANT
FREE
HEALTH BULLETINS

Let Patricia Bragg send you, your relatives and friends the latest News Bulletins on Health and Nutrition Discoveries. These are sent periodically. Please enclose two stamps for each U.S.A. name listed. Foreign listings send international postal reply coupons. Please print or type addresses, thank you.

HEALTH SCIENCE Box 7, Santa Barbara, California 93102 U.S.A.

●
_____
Name

_____ ( ) _____
Address   Phone

_____
City   State   Zip Code

●
_____
Name

_____ ( ) _____
Address   Phone

_____
City   State   Zip Code

●
_____
Name

_____ ( ) _____
Address   Phone

_____
City   State   Zip Code

●
_____
Name

_____ ( ) _____
Address   Phone

_____
City   State   Zip Code

●
_____
Name

_____ ( ) _____
Address   Phone

_____
City   State   Zip Code

BRAGG ALL NATURAL LIQUID AMINOS
Order Form

Delicious, Healthy Alternative to Tamari-Soy Sauce

BRAGG LIQUID AMINOS — Nutrition you need . . . taste you will love . . . a family favorite or over 65 years. A source of delicious, nutritious life-renewing protein. Add to casseroles, soups, sauces, gravies, potatoes, popcorn, and vegetables. An ideal "pick-me-up" broth at work, home or the gym. Tastes better than Tamari & Soy Sauce. Start today and add more amino Acids for healthy living to your daily diet – the easy BRAGG LIQUID AMINOS Way! Try a DASH for NEW TASTE DELIGHTS! **PROVEN AND ENJOYED BY MILLIONS.**

DELICIOUS, NUTRITIOUS FAMILY FAVORITE FOR OVER 65 YEARS!

Dash of Bragg Aminos brings New Taste Delights to Season:
- Salads & Dressings ■ Soups
- Vegies ■ Rice & Beans ■ Tofu
- Tempeh ■ Wok & Stir-frys
- Casseroles ■ Potatoes ■ Meats
- Poultry ■ Fish ■ Popcorn
- Gravies & Sauces ■ Macrobiotics

Pure Soybeans and Pure Water Only
- No Added Sodium
- No Coloring Agents
- No Preservatives
- Not Fermented
- No Chemicals
- No Additives

BRAGG LIQUID AMINOS

ZE	PRICE	SHIPPING		AMT.	$ TOTAL
oz.	$ 3.95 ea.	Please add $3.00 for 1st bottle / $1.50 for each additional bottle			
oz.	6.45 ea.	Please add $3.90 for 1st bottle / $1.90 for each additional bottle			
oz.	47.40	case/ 12 bottles	add $7 per case		
oz.	77.40	case/ 12 bottles	add $10 per case		

Please Specify: (U.S. Funds Only) — Prices subject to change without notice.
- [ ] Check [ ] Money Order
- [ ] Cash [ ] Credit Card

Total for Aminos	
Shipping	
Total Amount Enclosed	$

Charge My Order To: [ ] Visa [ ] MasterCard

Card Expires: MO. YR.

MasterCard InterBank No. (No. above name on card)

Credit Card Number:

Signature: _____

CREDIT CARD CUSTOMERS USE OUR FAST PHONE SERVICE: (805) 968-1020

In a hurry? Call (805) 968-1020. We can accept MasterCard or VISA phone orders only. Please prepare your order using this order form. It will speed your call and serve as your order record. Hours: 9am to 4pm Pacific Time, Monday to Thursday. FAX (805) 968-1001.

Mail to: HEALTH SCIENCE, Box 7, Santa Barbara, CA 93102 U.S.A.
(Please Print or Type) (Be sure to give street and house number to facilitate delivery)

ROF-8901

Name _____

Address _____ Apt. No. _____

City _____ State _____

Phone ( ) _____ Zip _____

Bragg Aminos – Taste You Love, Nutrition You Need

PAUL C. BRAGG N.D., Ph.D.

Life Extension Specialist ★ World Health Crusader
Lecturer and Advisor to Olympic Athletes, Royalty, and Stars

Originator of Health Food Stores - Now World-wide

For almost a Century, Living Proof that his
"Health and Fitness Way of Life" Works Wonders!

Paul C. Bragg can truly be considered the Grandfather of modern Natural Nutrition. This dynamic Crusader for worldwide health and fitness is responsible for more "firsts" in the history of modern Health Culture than any other individual. Here are a few of his incredible pioneering achievements that Americans now enjoy:

★ Bragg opened and named the first "Health Food Store" in America.
★ Bragg Crusades pioneered the first health lectures throughout America, inspiring followers to open health stores in cities across the land.
★ Bragg introduced pineapple juice and tomato juice to the American public.
★ He was the first to introduce and distribute honey nationwide.
★ He introduced Juice Therapy in America by importing the first hand-juicers.
★ Bragg pioneered Radio Health Programs from Hollywood three times daily.
★ Paul and Patricia pioneered a Health TV show from Hollywood to spread "Health and Happiness" ... the name of the show! It included exercises, health recipes, visual demonstrations, and guest appearances of famous, health-minded people.
★ He pioneered the first health foods and nationwide availability of herb teas, health beverages, seven-grain cereals and crackers, health cosmetics, health candies, vitamins, and minerals. He inspired others to follow and now thousands of health items are available worldwide.

Crippled by TB as a teenager, Bragg developed his own eating, breathing, and exercising program to rebuild his body into an ageless, tireless, painfree citadel of glowing, radiant health. He excelled in running, swimming, biking, progressive weight training, and mountain-climbing. He made an early pledge to God, in return for his renewed health, to spend the rest of his life showing others the road to health ... Paul Bragg made good his pledge.!

A living legend and beloved counselor to millions, Bragg was the inspiration and personal advisor on diet and fitness to top Olympic Stars from 4-time swimming Gold Medalist Murray Rose to 3-time track Gold Medalist Betty Cuthbert of Australia, his relative Don Bragg (pole-vaulting Gold Medalist), and countless others. Jack LaLanne, "TV's King of Fitness," says, "Bragg saved my life at age 14 when I attended the Bragg Crusade in Oakland, California." From the earliest days, Bragg was advisor to the greatest Hollywood Stars, and to giants of American Business. Gloria Swanson, Bob Cummings, J.C. Penney, Del Webb, and Conrad Hilton are just a few that he inspired to long, successful, healthy, active lives!

Dr. Bragg changed the lives of millions worldwide in all walks of life ... through his Health Crusades, Books, Tapes, and Radio, TV, and personal appearances.

HEALTH SCIENCE
Box 7, Santa Barbara, California 93102 U.S.A.